Global Sensitivity Analysis. The Primer

Global Sensitivity Analysis. The Primer

Andrea Saltelli, Marco Ratto,
Joint Research Centre of the European Commission, Ispra, Italy

Terry Andres
Department of Computer Science, University of Manitoba, Canada

Francesca Campolongo, Jessica Cariboni, Debora Gatelli, Michaela Saisana and Stefano Tarantola
Joint Research Centre of the European Commission, Ispra, Italy

John Wiley & Sons, Ltd

Copyright © 2008   John Wiley & Sons Ltd, The Atrium, Southern Gate, Chichester,
West Sussex PO19 8SQ, England
Telephone   (+44) 1243 779777

Email (for orders and customer service enquiries): cs-books@wiley.co.uk
Visit our Home Page on www.wiley.com

All Rights Reserved. No part of this publication may be reproduced, stored in a retrieval system or transmitted in any form or by any means, electronic, mechanical, photocopying, recording, scanning or otherwise, except under the terms of the Copyright, Designs and Patents Act 1988 or under the terms of a licence issued by the Copyright Licensing Agency Ltd, 90 Tottenham Court Road, London W1T 4LP, UK, without the permission in writing of the Publisher. Requests to the Publisher should be addressed to the Permissions Department, John Wiley & Sons Ltd, The Atrium, Southern Gate, Chichester, West Sussex PO19 8SQ, England, or emailed to permreq@wiley.co.uk, or faxed to (+44) 1243 770620.

This publication is designed to provide accurate and authoritative information in regard to the subject matter covered. It is sold on the understanding that the Publisher is not engaged in rendering professional services. If professional advice or other expert assistance is required, the services of a competent professional should be sought.

Other Wiley Editorial Offices

John Wiley & Sons Inc., 111 River Street, Hoboken, NJ 07030, USA

Jossey-Bass, 989 Market Street, San Francisco, CA 94103-1741, USA

Wiley-VCH Verlag GmbH, Boschstr. 12, D-69469 Weinheim, Germany

John Wiley & Sons Australia Ltd, 42 McDougall Street, Milton, Queensland 4064, Australia

John Wiley & Sons (Asia) Pte Ltd, 2 Clementi Loop #02-01, Jin Xing Distripark, Singapore 129809

John Wiley & Sons Canada Ltd, 6045 Freemont Blvd, Mississauga, ONT, L5R 4J3

Wiley also publishes its books in a variety of electronic formats. Some content that appears in print may not be available in electronic books.

Library of Congress Cataloging-in-Publication Data

Global sensitivity analysis. The Primer / Andrea Saltelli ... [et al.].
  p. cm.
  Includes bibliographical references and index.
  ISBN 978-0-470-05997-5 (cloth : acid-free paper)
  1. Sensitivity theory (Mathematics)   2. Global analysis (Mathematics)
  3. Mathematical models.   I. Saltelli, A. (Andrea), 1953–
  QA402.3.G557 2008
  003—dc22
                                                         2007045551

British Library Cataloguing in Publication Data

A catalogue record for this book is available from the British Library

ISBN 978-0-470-05997-5

Typeset in 10/12 Times by Integra Software Services Pvt. Ltd, Pondicherry, India

This book is dedicated to the Russian mathematician Ilya M. Sobol'

Contents

Preface		xi

1 Introduction to Sensitivity Analysis   1
- 1.1 Models and Sensitivity Analysis   1
  - 1.1.1 Definition   1
  - 1.1.2 Models   2
  - 1.1.3 Models and Uncertainty   3
  - 1.1.4 How to Set Up Uncertainty and Sensitivity Analyses   5
  - 1.1.5 Implications for Model Quality   9
- 1.2 Methods and Settings for Sensitivity Analysis – an Introduction   10
  - 1.2.1 Local versus Global   11
  - 1.2.2 A Test Model   12
  - 1.2.3 Scatterplots versus Derivatives   13
  - 1.2.4 Sigma-normalized Derivatives   15
  - 1.2.5 Monte Carlo and Linear Regression   16
  - 1.2.6 Conditional Variances – First Path   20
  - 1.2.7 Conditional Variances – Second Path   21
  - 1.2.8 Application to Model (1.3)   22
  - 1.2.9 A First Setting: 'Factor Prioritization'   24
  - 1.2.10 Nonadditive Models   25
  - 1.2.11 Higher-order Sensitivity Indices   29
  - 1.2.12 Total Effects   31
  - 1.2.13 A Second Setting: 'Factor Fixing'   33
  - 1.2.14 Rationale for Sensitivity Analysis   34
  - 1.2.15 Treating Sets   36
  - 1.2.16 Further Methods   37
  - 1.2.17 Elementary Effect Test   38
  - 1.2.18 Monte Carlo Filtering   39
- 1.3 Nonindependent Input Factors   41
- 1.4 Possible Pitfalls for a Sensitivity Analysis   41
- 1.5 Concluding Remarks   42

1.6	Exercises	44
1.7	Answers	44
1.8	Additional Exercises	50
1.9	Solutions to Additional Exercises	51

2 Experimental Designs — 53

2.1	Introduction	53
2.2	Dependency on a Single Parameter	55
2.3	Sensitivity Analysis of a Single Parameter	58
	2.3.1 Random Values	58
	2.3.2 Stratified Sampling	59
	2.3.3 Mean and Variance Estimates for Stratified Sampling	61
2.4	Sensitivity Analysis of Multiple Parameters	64
	2.4.1 Linear Models	65
	2.4.2 One-at-a-time (OAT) Sampling	66
	2.4.3 Limits on the Number of Influential Parameters	70
	2.4.4 Fractional Factorial Sampling	71
	2.4.5 Latin Hypercube Sampling	76
	2.4.6 Multivariate Stratified Sampling	80
	2.4.7 Quasi-random Sampling with Low-discrepancy Sequences	82
2.5	Group Sampling	89
2.6	Exercises	96
2.7	Exercise Solutions	99

3 Elementary Effects Method — 109

3.1	Introduction	109
3.2	The Elementary Effects Method	110
3.3	The Sampling Strategy and its Optimization	112
3.4	The Computation of the Sensitivity Measures	116
3.5	Working with Groups	121
3.6	The EE Method Step by Step	123
3.7	Conclusions	127
3.8	Exercises	128
3.9	Solutions	131

4 Variance-based Methods — 155

4.1	Different Tests for Different Settings	155
4.2	Why Variance?	157
4.3	Variance-based Methods. A Brief History	159
4.4	Interaction Effects	161
4.5	Total Effects	162
4.6	How to Compute the Sensitivity Indices	164

CONTENTS

4.7	FAST and Random Balance Designs	167
4.8	Putting the Method to Work: The Infection Dynamics Model	169
4.9	Caveats	174
4.10	Exercises	174

5 **Factor Mapping and Metamodelling** 183
  With Peter Young

5.1	Introduction	183
5.2	Monte Carlo Filtering (MCF)	184
	5.2.1 Implementation of Monte Carlo Filtering	185
	5.2.2 Pros and Cons	187
	5.2.3 Exercises	189
	5.2.4 Solutions	190
	5.2.5 Examples	200
5.3	Metamodelling and the High-Dimensional Model Representation	212
	5.3.1 Estimating HDMRs and Metamodels	214
	5.3.2 A Simple Example	224
	5.3.3 Another Simple Example	227
	5.3.4 Exercises	229
	5.3.5 Solutions to Exercises	231
5.4	Conclusions	235

6 **Sensitivity Analysis: From Theory to Practice** 237

6.1	Example 1: A Composite Indicator	238
	6.1.1 Setting the Problem	238
	6.1.2 A Composite Indicator Measuring Countries' Performance in Environmental Sustainability	239
	6.1.3 Selecting the Sensitivity Analysis Method	241
	6.1.4 The Sensitivity Analysis Experiment and Results	242
	6.1.5 Conclusions	252
6.2	Example 2: Importance of Jumps in Pricing Options	253
	6.2.1 Setting the Problem	253
	6.2.2 The Heston Stochastic Volatility Model with Jumps	255
	6.2.3 Selecting a Suitable Sensitivity Analysis Method	258
	6.2.4 The Sensitivity Analysis Experiment and Results	258
	6.2.5 Conclusions	261
6.3	Example 3: A Chemical Reactor	262
	6.3.1 Setting the Problem	262
	6.3.2 Thermal Runaway Analysis of a Batch Reactor	263
	6.3.3 Selecting the Sensitivity Analysis Method	266

                6.3.4   The Sensitivity Analysis Experiment and
                        Results   266
                6.3.5   Conclusions   269
        6.4   Example 4: A Mixed Uncertainty–Sensitivity Plot   270
                6.4.1   In Brief   270
        6.5   When to use What?   272

Afterword   277

Bibliography   279

Index   287

Preface

In the field of modelling it is easier to find academic papers, guidelines tailored to specific disciplines and handbooks of numerical simulation rather than plain textbooks of broad appeal. The various academic communities go about modelling largely independently of each other. Is this an indication that modelling is not a science but a craft, as argued by epistemologists? In other words, is it because it is impossible to define a single set of rules to encode natural or man-made systems into sets of mathematical rules called models?

If modelling is in fact characterized by such heterogeneity and lack of systematization, it might seem overly ambitious to offer a set of good practices of universal application in sensitivity analysis. Furthermore, if one looks at the available literature, in most instances 'sensitivities' are understood as derivatives of a particular output versus a particular input (such as elasticities in economics). This is not surprising, as contemporary researchers – like the authors of the present volume – are likely to have received more training in calculus than in Monte Carlo methods and to have seen more Jacobians and Hessians than Russian roulettes. A minority of sensitivity analysis practitioners (mostly in statistics, risk analysis and reliability) actively use importance measures such as those described in this book, whereby the influence of factors on outputs is assessed by looking at the entire input space rather than at a point in that space. Slowly these methods are finding their way into more recent modelling guidelines in other disciplines (see, for example, those of the Environmental Protection Agency in the USA, EPA, 2001). The purpose of this book is to offer to students an easy-to-read manual for sensitivity analysis covering importance measures and to show how these global methods may help to produce more robust or parsimonious models as well as to make models more defensible in the face of scientific or technical controversy.

1
Introduction to Sensitivity Analysis

1.1 MODELS AND SENSITIVITY ANALYSIS

WHAT IS A MODEL? WHAT MODEL INPUT IS CONSIDERED IN A SENSITIVITY ANALYSIS? WHAT IS THE ROLE OF UNCERTAINTY AND SENSITIVITY ANALYSES IN MODEL BUILDING? MAIN APPROACHES TO THE PROPAGATION OF UNCERTAINTY WITHIN AND ACROSS MODELS. IMPLICATIONS FOR MODEL QUALITY.

1.1.1 Definition

A possible definition of sensitivity analysis is the following: *The study of how uncertainty in the output of a model (numerical or otherwise) can be apportioned to different sources of uncertainty in the model input* (Saltelli et al., 2004). A related practice is 'uncertainty analysis', which focuses rather on *quantifying* uncertainty in model output. Ideally, uncertainty and sensitivity analyses should be run in tandem, with uncertainty analysis preceding in current practice.

For this definition of sensitivity analysis to be of use, it must first be made clear what is meant here by 'model', numerical or otherwise, as well as by the terms 'input' and 'output' which will be used throughout this book.

Global Sensitivity Analysis. The Primer A. Saltelli, M. Ratto, T. Andres, F. Campolongo, J. Cariboni, D. Gatelli, M. Saisana and S. Tarantola © 2008 John Wiley & Sons, Ltd

1.1.2 Models

A view of modelling that may help to illustrate the role of sensitivity analysis in the scientific process is offered in Figure 1.1, taken from the work of biologist Robert Rosen (1991) (see also Saltelli *et al.*, 2000, pp. 3–4). On the left in Rosen's diagram we have the 'world', that is the system which forms the subject of our investigation. We have reason to believe that the system, whether natural or artificial, is governed by rules which we have the ambition to uncover, or to use to our advantage. To this end we craft or hypothesize a set of structures in a model (depicted on the right-hand side of the figure). For example, a hypothesized growth mechanism for a species contained in the world can be translated into a differential equation in a model. While our species continues growing and dying quietly in the world, following the forces of its own systemic causality (which we aim to understand), our differential equation can be solved using the rules of mathematical calculus. The intuition of Rosen is that while the species in the world obeys rules, and the differential equation in the model has 'rules' as well, whether formal or mathematical, no 'rule' whatsoever can dictate how one should map the hypothesized rules in the world onto the rules in the model. In the words of Rosen, while the world and the model are each internally 'entailed', nothing entails the world with the model. Among the reasons for this paradox is the fact that the portion of the world captured by the model is an arbitrary 'enclosure' of an otherwise open, interconnected system.[1] This is the case when the world is part of a natural system, the main concern of Rosen's inquiry. Yet experience has shown that even when the world is indeed a well-defined and closed system, for instance an artefact, an artificial device or a piece of machinery, different

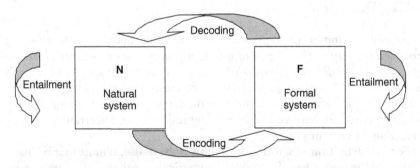

Figure 1.1 Modelling after Rosen (1991)

---

[1] Even more so when the purpose of a model is to learn about the nonobservable parts of a system.

modellers can generate different nonequivalent descriptions of it, that is, models whose outputs are compatible with the same set of observations but whose structures are not reconcilable with one another.

While this may be disturbing to a student accustomed to the beauty and apparent self-evidence of physical laws, practitioners of modelling have come to live with the rather unpleasant reality that more than one model may be compatible with the same set of data or evidence. Some have gone so far as to coin a word for this paradox: equifinality – Beven (1993, 2001), see also Saltelli *et al.* (2004, pp. 173–178) – meaning that different models can lead to the same end. Others refer to the phenomenon as model indeterminacy.

Since Galileo's time scientists have had to deal with the limited capacity of the human mind to create useful maps of 'world' into 'model'. The emergence of 'laws' can be seen in this context as the painful process of simplification, separation and identification which leads to a model of uncharacteristic simplicity and beauty.

1.1.3 Models and Uncertainty

What makes modelling and scientific inquiry in general so painful is uncertainty. Uncertainty is not an accident of the scientific method, but its substance.[2]

Modellers and philosophers of science have debated the issue of model indeterminacy at length (Oreskes *et al.*, 1994). Most modellers today would probably agree that a model cannot be validated, in the sense of 'be proven true'. Rather, it is more defensible and correct to say that a model has been extensively corroborated, meaning by this that the model has survived a series of tests – be they formal, of internal consistency, or relative to the model's capacity to explain or predict the 'world' in a convincing and parsimonious way.

When models fail publicly, the ensuing controversy can be devastating for the scientific parties involved.[3] Models are often used in highly polarized contexts and uncertainty may be used instrumentally. 'All parties deal with environmental information in a selective way, or even manipulate it', observed a Dutch environmental scientist (In 't Veld, 2000). Fabricated

---

[2] 'That is what we meant by science. That both question and answer are tied up with uncertainty, and that they are painful. But that there is no way around them. And that you hide nothing; instead, everything is brought out into the open' (Høeg, 1995).
[3] For the modelling credibility crisis in the Netherlands' RIVM Laboratories see Van der Sluijs (2002). See also Mac Lane (1988) for another example.

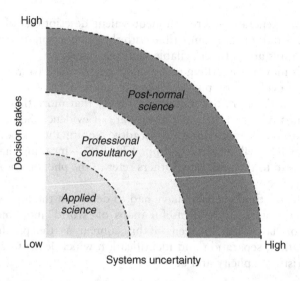

Figure 1.2 Uncertainty/stakes diagram after Funtowicz and Ravetz (1990)

uncertainty is a common concern in relation to important disputes over health or the environment (Michaels, 2005).

In short, models are part of the scientific method and hence subject to epistemological debate. A way of framing present-day debate on the scientific method is offered by Post-Normal Science (PNS, see Figure 1.2 and Funtowicz and Ravetz, 1990, 1993; Funtowicz *et al.*, 1996).

In PNS one distinguishes between three types of scientific production modes, depending on the system's uncertainties and the stakes involved. Applying this to modelling, different requirements and practices pertain:

- In applied science, when a model is written and employed within a closed consortium of experts who are the sole users of the model, e.g. when this is used to solve a circumscribed chemical kinetics problem;
- In 'consultancy' when the model is more likely to be scrutinized, e.g. as part of a cost–benefit analysis for the construction of a new road or bridge that will affect a community;
- When computing climate sensitivity in the context of global change. In this latter case we are in the domain of PNS, where science (and its models) is called on to provide evidence under circumstances of conflicting stakes and beliefs.

Like scientific theories, models may be given pedigrees which help us to judge their quality. Pedigrees take account of past usage of the model, status of its proponents, degree of acceptance by peers and so on (Van der Sluijs, 2002; Craye *et al.*, 2005). In pedigrees, model quality is more closely associated with 'fitness for purpose' – that is, with a specific purpose – than with the model's intrinsic fabric.

A post-normal view of the modes of scientific production in relation to policy is given in Funtowicz (2004). Models as metaphors are discussed in Ravetz (2006).

1.1.4 How to Set Up Uncertainty and Sensitivity Analyses

As mentioned at the beginning of the chapter, our definition of sensitivity analysis involves models, model input and model output. We now try to define model input in relation to the nature and purpose of the model, as well as to the set-up of the uncertainty and sensitivity analyses. A model can be:

- *Diagnostic or prognostic.* In other words, we try to distinguish between models used to understand a law and models used to predict the behaviour of a system given a supposedly understood law. Models can thus range from wild speculations used to play what-if games (e.g. models for the existence of extraterrestrial intelligence) to models which can be considered accurate and trusted predictors of a system (e.g. a control system for a chemical plant).
- *Data-driven or law-driven.* A law-driven model tries to put together accepted laws which have been attributed to the system, in order to predict its behaviour. For example, we use Darcy's and Ficks' laws to understand the motion of a solute in water flowing through a porous medium. A data-driven model tries to treat the solute as a signal and to derive its properties statistically. Advocates of data-driven models like to point out that these can be built so as to be parsimonious, i.e. to describe reality with a minimum of adjustable parameters (Young et al., 1996). Law-driven models, by contrast, are customarily overparametrized, as they may include more relevant laws than the amount of available data would support. For the same reason, law-driven models may have a greater capacity to describe the system under unobserved circumstances, while data-driven models tend to adhere to the behaviour associated with the data used in their estimation. Statistical models (such as hierarchical or multilevel models) are another example of data-driven models.

Many other categorizations of models are possible,[4] and the definition of model input depends on the particular model under study. For the purpose

---

[4] Bell et al. (1988) distinguish between formal (axiomatic), descriptive and normative models (rules an agent should follow to reach a target). The examples in this book are descriptive models.

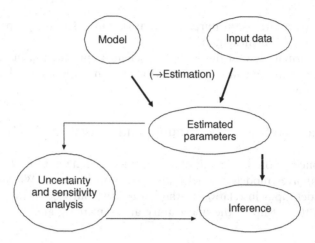

Figure 1.3 Parametric bootstrap version of uncertainty and sensitivity analyses

of uncertainty and sensitivity analyses we could liberally classify as input everything that can drive a variation in the output of the model.

Consider the scheme in Figure 1.3. Here we have observations (assumed error-free for simplicity's sake) and a model whose parameters are estimated from the data. Estimation can take different courses. Usually it is achieved by minimizing, e.g. by least squares, some measure of distance between the model's prediction and the data. At the end of the estimation step, 'best' parameter values as well as their errors are known. At this point we might consider the model 'true' and run an uncertainty analysis by propagating the uncertainty in the parameters through the model, all the way to the model output. In this case the estimated parameters become our factors.

One way of doing this is through Monte Carlo analysis, in which we look at the distribution functions of the input parameters, as derived from the estimation. For example, we may have the following scheme:

- We start from a factor $\alpha \sim N(\bar{\alpha}, \sigma_\alpha)$, which reads: after estimation α is known to be normally distributed with mean $\bar{\alpha}$ and standard deviation σ_α.
- Likewise for factors β, γ and so on. Contrary to what logic would suggest, and for the sake of simplicity, we assume that the factors are independent of each other. This issue is discussed later in the chapter.
- For each of these factors, we draw a sample from the respective distributions, i.e. we produce a set of row vectors $(\alpha^{(j)}, \beta^{(j)}, \ldots)$ with $j = 1, 2, \ldots, N$ in such a way that $(\alpha^{(1)}, \alpha^{(2)}, \ldots, \alpha^{(N)})$ is a sample

from $N(\bar{\alpha}, \sigma_\alpha)$ and likewise for the distribution function of the other factors.

$$\begin{bmatrix} \alpha^{(1)} & \beta^{(1)} & \gamma^{(1)} & \dots \\ \alpha^{(2)} & \beta^{(2)} & \gamma^{(2)} & \dots \\ \dots & \dots & \dots & \dots \\ \alpha^{(N-1)} & \beta^{(N-1)} & \gamma^{(N-1)} & \dots \\ \alpha^{(N)} & \beta^{(N)} & \gamma^{(N)} & \dots \end{bmatrix} \quad (1.1)$$

- We can then compute ('run' is the conventional term) the model for all vectors $(\alpha^{(j)}, \beta^{(j)}, \dots)$ thereby producing a set of N values of a model output Y_j.[5]

$$\begin{bmatrix} y^{(1)} \\ y^{(2)} \\ \dots \\ y^{(N-1)} \\ y^{(N)} \end{bmatrix} \quad (1.2)$$

These steps constitute our uncertainty analysis. From these we can compute the average output, its standard deviation, the quantiles of its distribution, confidence bounds, plot the distribution itself and so on. It is clear that in this analysis, sometimes called a 'parametric bootstrap',[6] our inputs are the model's parameters. Having performed this uncertainty analysis we can then move on to a sensitivity analysis, in order to determine which of the input parameters are more important in influencing the uncertainty in the model output. However, we defer this step in order to continue our discussion of model input.

Note that for the purpose of the uncertainty analysis just described we consider as relevant inputs only our estimated parameters. All other types of information fed into the model, e.g. the observations, physical or mathematical constants, internal model variables (e.g. number of grid points if the model needs a mesh), are disregarded – that is, we do not allow them to vary and hence they cannot cause variation in the output.

In Figure 1.4 we have played the uncertainty analysis game differently by sampling the observations rather than the parameters. We have a limited set of observations, and we are aware that different subsets of these could

---

[5] Note that this model output Y_j may be different from the model output used in the estimation step.
[6] Bootstrapping is the process of repeatedly sampling 'with replacements'. For example, if we want to estimate the average sum of three Bingo chips, we could do this by extracting three random chips from the Bingo bag, computing their average, putting the chips back into the bag and extracting again. With a sufficiently large number of extractions we could determine the average sum being sought, and this strategy would be called a bootstrap of the Bingo chips.

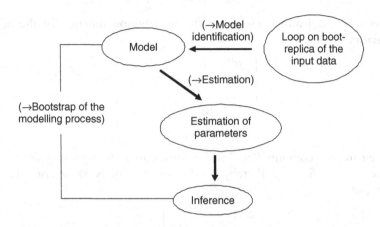

Figure 1.4 Bootstrapping of the modelling process (Chatfield, 1993)

potentially lead us to try one model rather than another to fit the data. What we can do in order to be fair to the data is to select a subset of the observations, choose a model based on these data using a pre-established model selection rule, estimate the corresponding parameters using the same sampled data, and run the model to compute Y_j. We have drawn the sample with replacement, and we can now repeat the process, identifying a potentially different (or indeed the same) model, estimating the parameters (which may differ in number from those of the previous run if the model is different), and so on for a total of N times, until we yield our desired sample for the uncertainty analysis. This approach can be called 'bootstrapping of the modelling process' (Chatfield, 1993).

The input for this uncertainty analysis is the data which have been bootstrapped, since we have assumed that all the rest (from model selection to parametric estimation) is done automatically given the data and hence adds no variation to model output.

Finally in Figure 1.5 we compare a set of plausible models with the data. Using Bayesian analysis it is possible to derive posterior probabilities for the models as well as distributions of the related parameters (Saltelli et al., 2004). Once this model update and parameter estimation step is complete, a model averaging can be used in uncertainty analysis. This is done by propagating the uncertainty through the system by sampling both the model and the parameters according to their distributions, to produce a sample of model outcome Y_j. This procedure is known as Bayesian model averaging,[7] and the inputs in this case are both models and parameters, or

---

[7] For a thorough account of this approach see Kass and Raftery (1995) and Hoeting et al. (1999). See Saltelli et al. (2004, pp. 151–192) for related sensitivity issues.

MODELS AND SENSITIVITY ANALYSIS

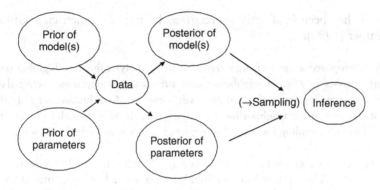

Figure 1.5 Bayesian model averaging

more precisely the probabilities of the different model representations and the distributions of the parameters. In a Monte Carlo framework, a trigger variable would be sampled to select a model according to its posterior probability, while the parameters would also be sampled and the model outcome determined. A sensitivity analysis could be executed at this point, and a question that it might address is the following: how much of the uncertainty is due to the model selection and how much to the estimation of the parameters?

1.1.5 Implications for Model Quality

The superficial illustration given above of approaches to uncertainty and sensitivity analyses has shown that what constitutes an input for the analysis depends upon how the analysis is set up. The input is that which is allowed to vary in order to study its effect on the output. A sensitivity analysis will in turn instruct the modellers as to the relative importance of the inputs in determining the output. An obvious consequence of this is that the modeller will remain ignorant of the importance of those variables which have been kept fixed. This is of course a hazard for the modeller, as a variable deemed noninfluential and kept fixed could haunt the results of the analysis at a later stage. For example, it would be unfortunate for the modeller to discover a posteriori that the mesh size had been too large, and that the number of grid points had had a dramatic effect on the model output.

It seems, therefore, that one should be as careful and objective as possible in deciding on the input for uncertainty and sensitivity analyses. Clearly, the more variables we promote to the rank of input, and allow to vary, the greater the variance to be expected in the model prediction. This could lead to a situation in which we discover that, having incorporated all uncertainties, the model prediction varies so wildly as to be of no practical use. This trade-off has been brilliantly summarized by the econometrician Edward E. Leamer (1990):

> I have proposed a form of organized sensitivity analysis that I call 'global sensitivity analysis' in which a neighborhood of alternative assumptions is selected and the corresponding interval of inferences is identified. Conclusions are judged to be sturdy only if the neighborhood of assumptions is wide enough to be credible and the corresponding interval of inferences is narrow enough to be useful.

Note Leamer's emphasis on the need for 'credibility' in the selection of assumptions. The easiest way to invalidate a model is to demonstrate it fragile with respect to shaky assumptions. Note, however, that the trade-off may not be as dramatic as one might expect, and that increasing the number of input factors does not necessarily lead to an increased variance in model output. Practitioners have recorded that in most uncertainty and sensitivity analyses the input factors' importance is distributed similarly to wealth in nations, with a few factors creating almost all the uncertainty and the majority making only a negligible contribution. Hence, if the 'key' factors have been judiciously chosen, adding further variables to the analysis may add to its completeness and defensibility without adversely increasing the variance in the output.

As mentioned, the quality of a model is largely a function of its fitness for purpose. If modelling is a craft and models cannot be proven true (because of the pervasive nature of uncertainty and the difficulty of separating observation from observer and facts from values),[8] then the modeller has a moral obligation, and indeed it is in the modeller's own practical interest, to be as rigorous as possible when assessing the robustness of model inference. Doing so should produce better and more parsimonious models, and will strengthen the analyst's defence of the results in the case of scientific controversy or public policy debate.

1.2 METHODS AND SETTINGS FOR SENSITIVITY ANALYSIS – AN INTRODUCTION

WHAT METHODS ARE AVAILABLE? HOW CAN A PARTICULAR METHOD BE RELATED TO A PROBLEM-SPECIFIC QUESTION? HOW CAN WE DEFINE A FACTOR'S IMPORTANCE UNAMBIGUOUSLY? SUGGESTED PRACTICES.

---

[8] 'Values' here mean ethical judgements. Cases in which the separation of facts and values becomes arduous are many, e.g. when models try to assess the impact of the adoption of new technologies, the relevance of environmental threats, distributional issues in economics and so on.

METHODS AND SETTINGS FOR SENSITIVITY ANALYSIS 11

1.2.1 Local versus Global

As we shall learn in the following chapters, sensitivity analysis can serve a number of useful purposes in the economy of modelling. It can surprise the analyst, uncover technical errors in the model, identify critical regions in the space of the inputs, establish priorities for research, simplify models and defend against falsifications of the analysis. In the context of models used for policy assessment, sensitivity analysis can verify whether policy options can be distinguished from one another given the uncertainties in the system, and so on. What methods would one choose to perform sensitivity analysis for any or all of the above?

It is not by chance that most of the sensitivity analyses met in the literature are based on derivatives. Indeed the derivative $\partial Y_j / \partial X_i$ of an output Y_j versus an input X_i can be thought of as a mathematical definition of the sensitivity of Y_j versus X_i.

Sometimes computer programs that implement complex physical, chemical or genetic models are augmented by special routines that allow the efficient computation of large arrays of system derivatives, which are subsequently used for model calibration, model reduction or verification and model inversion (Rabitz, 1989; Turanyi, 1990; Varma *et al.*, 1999; Cacuci, 2003; Saltelli *et al.*, 2000, pp. 81–101).

The derivative-based approach has the attraction of being very efficient in computer time. The model needs to be executed few times compared to the dimension of the array of derivatives to be computed. However, it is inefficient in terms of the analyst's time. One has to intervene in the computer program, inserting ad hoc coding, to perform this operation efficiently. Yet the fatal limitation of a derivative-based approach is that it is unwarranted when the model input is uncertain and when the model is of unknown linearity. In other words, derivatives are only informative at the base point where they are computed and do not provide for an exploration of the rest of the space of the input factors. This would matter relatively little for linear systems, in which the property at a point away from the baseline can be computed quickly by linear extrapolation using first-order point derivatives, but it would matter greatly for nonlinear ones. The focus of this book is on quantitative uncertainty and sensitivity analysis in the presence of uncertain inputs. We shall therefore make use of methods based on exploring the space of the input factors, based on the consideration that a handful of data points judiciously thrown into that space is far more effective, in the sense of being informative and robust, than estimating derivatives at a single data point in the centre of the space. In this book, when we use derivatives, or rather incremental ratios such as $(Y_j(X_i + \Delta X_i) - Y_j(X_i))/\Delta X_i$, we will normally compute them at a set of different points in the space of the input factors, in order to obtain an average response of Y_j when moving a factor X_i of a step ΔX_i at different points in the input space, i.e. for different values of $\mathbf{X}_{\sim i}$.[9]

However, in order to introduce the methods of sensitivity analysis, we shall start from derivatives, taking a very simple test case.

1.2.2 A Test Model

Imagine the model has a linear error-free form

$$Y = \sum_{i=1}^{r} \Omega_i Z_i \qquad (1.3)$$

where the input factors are $\mathbf{X} = (\Omega_1, \Omega_2, \ldots \Omega_r, Z_1, Z_2, \ldots Z_r)$.

We have dropped the subscript j of the model output Y for simplicity. Model equation (1.3) has just a single output variable. Let us assume first that the Ω's are fixed coefficients, so that the true (active) factors for model (1.3) are just the $Z_1, Z_2, \ldots Z_r$. Y could be a composite indicator, for example a sustainability index or a greenhouse gas emission index, in which the Ω's are the weights attached by experts to the individual Z-variables. For the sake of the example we consider the weights fixed, while the individual variables have been characterized as independent and distributed normally with mean zero, i.e.

$$Z_i \sim N(0, \sigma_{Z_i}) \quad i = 1, 2, \ldots, r. \qquad (1.4)$$

If the model were indeed a composite indicator with 'standardized' variables[10] all σ_{Z_i}'s would be equally one.

It is easy to verify (see the Exercises) that, given the Equations (1.3, and 1.4), Y will also be normally distributed with parameters

$$\bar{y} = \sum_{i=1}^{r} \Omega_i \bar{z}_i \qquad (1.5)$$

$$\sigma_Y = \sqrt{\sum_{i=1}^{r} \Omega_i^2 \sigma_{Z_i}^2}. \qquad (1.6)$$

---

[9] Here, and in the following, $\mathbf{X}_{\sim i}$ denotes the vector of all factors but X_i.
[10] Standardization of a variable is achieved by subtracting from the variable its sample mean and dividing the result by its standard deviation.

For the sake of the example we would also like to assume that for this particular index the variables have been ordered from the less uncertain to the most uncertain, i.e.

$$\sigma_{Z_1} < \sigma_{Z_2} < \ldots < \sigma_{Z_r},$$

and that the weights Ω's are all equal and constant:

$$\Omega_1 = \Omega_2 = \ldots = \Omega_r = \text{constant}. \tag{1.7}$$

1.2.3 Scatterplots versus Derivatives

Figure 1.6 shows the scatterplots Y, Z_i that we obtain by performing a Monte Carlo experiment with our model. As already mentioned (and described in more detail in Chapter 2), Monte Carlo methods are based on sampling factors' values from their distribution. In most cases factors are assumed independent so that the samples are taken from the marginal distribution of each factor. An input sample is thus produced:

$$\mathbf{M} = \begin{bmatrix} z_1^{(1)} & z_2^{(1)} & \ldots & z_r^{(1)} \\ z_1^{(2)} & z_2^{(2)} & \ldots & z_r^{(2)} \\ \ldots & \ldots & \ldots & \ldots \\ z_1^{(N-1)} & z_2^{(N-1)} & \ldots & z_r^{(N-1)} \\ z_1^{(N)} & z_2^{(N)} & \ldots & z_r^{(N)} \end{bmatrix} \tag{1.8}$$

Computing Y for each row of matrix (1.8) using model (1.3) produces the desired output vector

$$\mathbf{Y} = \begin{bmatrix} y^{(1)} \\ y^{(2)} \\ \ldots \\ y^{(N-1)} \\ y^{(N)} \end{bmatrix} \tag{1.9}$$

where $y^{(1)}$ is the value obtained by running Equation (1.3) with the input given by the row vector $z_1^{(1)}, z_2^{(1)}, \ldots, z_r^{(1)}$, and so on for the other rows of matrix (1.8).

With this sample of model input and output one can produce r scatterplots by projecting in turn the N values of the selected output Y (assumed here to be a scalar) against the N values of each of the r input factors. These scatterplots can be used to investigate the behaviour of models.

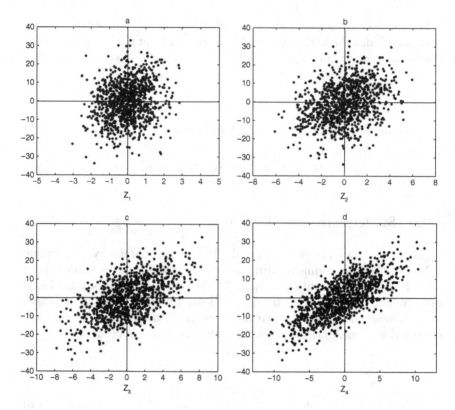

Figure 1.6 Scatterplots of Y versus $Z_1, \ldots, Z_4$. Which is the most influential factor? One can compare occupancy of quadrants I and III versus that of II and IV to decide where the positive linear relationship is stronger

The scatterplots show that Y is more sensitive to Z_4 than it is to Z_3, and that the ordering of the input factors by their influence on Y is

$$Z_4 > Z_3 > Z_2 > Z_1. \qquad (1.10)$$

Such a conclusion can be drawn from Figure 1.6, as there is more shape (or a better pattern) in the plot for Z_4 than for Z_3, and so on.

However, if we used the straightforward derivative of Y versus Z_i for the sensitivity analysis, i.e. if we decided upon the relative importance of the Z_i's using the measure

$$S^p_{Z_i} = \frac{\partial Y}{\partial Z_i}, \qquad (1.11)$$

which gives $S^p_{Z_i} = \Omega_i$ for Equation (1.3), we would have to conclude that all factors are equally important, based on Equation (1.7), irrespective of the values in σ. This is clearly not reasonable. Note that we have used the superscript p for 'partial derivative' in Equation (1.11), and that the derivative is nonnormalized, i.e. it is based on the raw values of both input and output. Note also that the scatterplots in Figure 1.6 are more convincing than formula (1.11) as a sensitivity analysis tool. This is a rather general conclusion. Input/output scatterplots are in general a very simple and informative way of running a sensitivity analysis – we will use them often in this book, since they can provide an immediate visual depiction of the relative importance of the factors. For example, a scatterplot with little 'shape', e.g. plot Z_1 in Figure 1.6, which presents a rather uniform cloud of points over the range of the input factor on the abscissa, is an almost sure sign that the parameter is less influential than factor Z_4. We say 'almost' because there are instances in which a bidimensional scatterplot can be deceiving, leading to type II errors (nonidentification of an influential factor).[11] These are, however, very special cases, see Saltelli et al. (2004, pp. 160–161).

Most sensitivity analysis measures developed by practitioners aim to preserve the rich information provided by scatterplots in condensed format. The challenge for sensitivity analysis, in situations with many input factors, is how to rank the factors rapidly and automatically without having to look at many separate scatterplots. Another problem with scatterplots is that some uncertain factors might be sets, that is, groups of factors, and while compact sensitivity measures can be defined for sets, the sensitivities of sets cannot be visualized via simple two-dimensional scatterplots.[12]

1.2.4 Sigma-normalized Derivatives

Can we improve Equation (1.11) in such a way as to obtain a sensitivity measure that would rank the input factors consistently with Figure 1.6? A good possibility is

$$S_{Z_i}^{\sigma} = \frac{\sigma_{Z_i} \partial Y}{\sigma_Y \partial Z_i}. \tag{1.12}$$

---

[11] In sensitivity analysis, we refer to type I error when erroneously defining as important a noninfluential factor. Type II error occurs when we classify an important factor as noninfluential. It is nowadays common practice in modelling to include a third type of error: type III. This is typically a framing error, where right answers are sought for the wrong question. Sensitivity analysis is unable to help against type III errors. To make an example, if the range of plausible values for a factor taken as input for a sensitivity analysis is totally off the mark, the result of the sensitivity analysis will be of little help.
[12] In fact, one can force multidimensional scatterplots into a bidimensional plane by scanning the space of the input factors with a search curve. See Chapter 5.

This derivative is normalized by the input–output standard deviations (hence the σ in the superscript). Applied to model (1.3) this would give $S_{Z_i}^{\sigma} = (\sigma_{Z_i}/\sigma_Y)\Omega_i$. Squaring this and comparing with the square of $\sigma_Y = \sqrt{\sum_{i=1}^{r} \Omega_i^2 \sigma_{Z_i}^2}$ (Equation (1.6) above) we obtain

$$\sigma_Y^2 = \sum_{i=1}^{r} \Omega_i^2 \sigma_{Z_i}^2 \quad \text{and} \quad \left(S_{Z_i}^{\sigma}\right)^2 = \left(\frac{\sigma_{Z_i}}{\sigma_Y}\Omega_i\right)^2 \quad (1.13)$$

which gives $\sigma_Y^2 = \sigma_Y^2 \sum_{i=1}^{r} \left(S_{Z_i}^{\sigma}\right)^2$, and finally

$$\sum_{i=1}^{r} \left(S_{Z_i}^{\sigma}\right)^2 = 1. \quad (1.14)$$

Measure (1.12) is more convincing than measure (1.11), see Table 1.1: first, because the relative ordering of the Z_i's now depends on both vectors, σ and Ω, just as it should; and second, because the sensitivity measures are neatly normalized to one.

Note that Equation (1.12) is a measure recommended for sensitivity analysis by a guideline of the Intergovernmental Panel for Climate Change (IPCC) (1999, 2000).

1.2.5 Monte Carlo and Linear Regression

Let us return briefly to the scatterplots of Figure 1.6. As mentioned, these are the result of a Monte Carlo simulation in which a matrix such as

$$\mathbf{M} = \begin{bmatrix} z_1^{(1)} & z_2^{(1)} & \ldots & z_r^{(1)} \\ z_1^{(2)} & z_2^{(2)} & \ldots & z_r^{(2)} \\ \ldots & \ldots & \ldots & \ldots \\ z_1^{(N-1)} & z_2^{(N-1)} & \ldots & z_r^{(N-1)} \\ z_1^{(N)} & z_2^{(N)} & \ldots & z_r^{(N)} \end{bmatrix} \quad (1.15)$$

Table 1.1 Derivatives and normalized derivatives for the model (1.3, 1.4), where $r = 4$, $\Omega = (2, 2, 2, 2)$ and $\sigma = (1, 2, 3, 4)$

	$S_{Z_i}^{a}$	$\left(S_{Z_i}^{\sigma}\right)^2$
Z_1	2	0.036
Z_2	2	0.14
Z_3	2	0.31
Z_4	2	0.56

METHODS AND SETTINGS FOR SENSITIVITY ANALYSIS

has been fed into model (1.3) to produce the desired output vector

$$\mathbf{Y} = \begin{bmatrix} y^{(1)} \\ y^{(2)} \\ \ldots \\ y^{(N-1)} \\ y^{(N)} \end{bmatrix} \qquad (1.16)$$

where $y^{(1)}$ is the value obtained running Equation (1.3) with the input given by the row vector $z_1^{(1)}, z_2^{(1)}, \ldots, z_r^{(1)}$, and so on for the other rows of the matrix.

Note that N is the size of the Monte Carlo experiment ($N = 1000$ in Figure 1.6). N corresponds to the number of times we have computed Equation (1.3). In a sensitivity analysis experiment we shall have in general, instead of Equation (1.3), a computer program that calculates $\mathbf{Y}$. Running the program to obtain a vector as Equation (1.9) is customarily the most expensive part of the analysis in terms of computer time, as the model may be complicated, while the sensitivity analysis measures are easy to compute. Thus N is referred to as the cost of the analysis. Note that computer time is not to be confused with analysis time. A derivation of the factors' uncertainty distribution such as Equation (1.4) is in practice the most time-consuming and financially expensive part of an analysis, especially when this is based on formal elicitation of expert opinion (Helton *et al.*, 2006; see also Saltelli *et al.*, 2000, pp. 101–152).

Note also that care has to be taken so that each column

$$\begin{matrix} z_1^{(1)} \\ z_1^{(2)} \\ \ldots \\ z_1^{(N)} \end{matrix}$$

in matrix (1.8) is a sample from the respective distribution $Z_i \sim N(\bar{z}_i, \sigma_{Z_i})$. In general, and unless otherwise specified, we assume that the input factors are independent of each other, so that each one can be independently sampled from its marginal distributions (Equation (1.4) in the present examples).

As mentioned above, analysts would like to summarize the results in plots such as Figure 1.6 with a single number per scatterplot. This is, after all, what a sensitivity measure is intended to do. The most popular method for this is to try a simple linear regression on the data of matrix (1.8) and vector (1.9), of the form

$$y(i) = b_0 + \sum_{j=1}^{r} b_{Z_j} z_j^{(i)}, \qquad (1.17)$$

where the coefficients b_0, b_{Z_i} are determined by least-square computation, based on the squared differences between the y-values produced by the regression (meta)model[13] and the actual model output produced by the Monte Carlo simulation. Because the points have been generated using a linear model, we expect that the linear regression will re-discover it, i.e. we would expect that $\hat{b}_0 \cong 0, \hat{b}_{Z_i} \cong \Omega_i, i = 1, 2, \ldots, r$, where the symbol $\cong$ means that this is what we would obtain if N were large enough and the hat denotes estimates as in standard usage.

Results for the points in Figure 1.6 ($N = 1000$) are given in Table 1.2.[14]

All available software for regression analysis will compute not only $\hat{b}_0, \hat{b}_{Z_i}$, but also their standardized equivalents $\hat{\beta}_{Z_i} = \hat{b}_{Z_i} \sigma_{Z_i}/\sigma_Y$. The β's are known as standardized regression coefficients (sometime indicated with their initial as SRC), and are in general more widely used than the raw regression coefficients b's. For our model (1.3), the regression coefficients, again for N tending to infinity, will tend to

$$\hat{\beta}_{Z_i} = \hat{b}_{Z_i} \sigma_{Z_i}/\sigma_Y \cong \Omega_i \sigma_{Z_i}/\sigma_Y. \quad (1.18)$$

Comparing this formula with that previously obtained for the σ-normalized derivatives, i.e. $S^\sigma_{Z_i} = (\sigma_{Z_i}/\sigma_Y)\Omega_i$, we can conclude that in the special case of our model (1.3) the two measures of sensitivity coincide:

$$\hat{\beta}_{Z_i} = S^\sigma_{Z_i} \quad \text{for model (1.3).} \quad (1.19)$$

Table 1.2 Linear regression coefficients and standardized coefficients for the model of Equations (1.3), (1.4), where $r = 4$, $\Omega = (4, 3, 2, 1)$ and $\sigma = (2, 2, 2, 2)$, $N = 1000$

	b	$\beta^2_{Z_i}$	S_{Z_i} Analytic (see Exercises)
Intercept	0		
Z_1	2	0.034	0.03
Z_2	2	0.14	0.13
Z_3	2	0.31	0.3
Z_4	2	0.53	0.53

---

[13] Metamodels are surrogate models which are built to substitute for computationally intensive simulation models. Metamodels can be built with a variety of strategies (e.g. simple linear regression as discussed above) and purposes (e.g. to perform a sensitivity analysis). See Chapter 5.

[14] The results in Table 1.2 have been obtained with a simple piece of software for regression analysis. Yet, as explained in the next chapter, given that our model (1.3, 1.4) is linear, and the model does not contain any error term, we could have computed exact (analytic) values of the regression coefficients using only five runs and then applying the Kramer formula for a system of five equations (runs) in the five unknowns $b_0, \ldots b_4$.

METHODS AND SETTINGS FOR SENSITIVITY ANALYSIS

As a result it will also be true for the β's that

$$\sum_{i=1}^{r} \left(\hat{\beta}_{Z_i}\right)^2 = 1 \tag{1.20}$$

when the model is linear.

The fact that the two measures coincide for our model can be generalized only to linear models and no further. If the model is nonlinear, the two measures will be different. Yet the β's will be a more robust and reliable measure of sensitivity even for nonlinear models. First of all, the β's are multidimensionally averaged measures. Unlike $S_{Z_i}^\sigma$, which is computed at the midpoint of the distribution of Z_i while keeping all other factors fixed at their midpoint, $\hat{\beta}_{Z_i}$ is the result of an exploration of the entire space of the input factors – the limit being in the dimension N of the sample. For small N and large r, however, the β's will be rather imprecise. Even in sensitivity analysis there is no such thing as a free meal, and one cannot expect to have explored a high-dimensionality space with a handful of points. Nevertheless a handful is better than just one. Statistical significance tests are available for the β's, so that the analysts can at least know the extent of the problem. Finally, by computing $\sum_{i=1}^{r} \left(\hat{\beta}_{Z_i}\right)^2$ or a related statistic, one will obtain a number, in general less than one, equal to the fraction of linearity of the model. More precisely, this number – known as the model coefficient of determination, and written as R_Y^2 – is equal to the fraction of the variance of the original data which come from our model (Equations 1.3, 1.4, in this case), which is explained by the regression model of Equation (1.17). Again, this fraction should be equal to one for our model; however, to give a different example, if R_Y^2 were instead to be of the order of 0.9, then the model would be 90% linear and one could use the β's for sensitivity analysis, at the risk of remaining ignorant of some 10% of the variance of the problem.[15]

Note that

$$\sum_{i=1}^{r} \left(\hat{\beta}_{Z_i}\right)^2 = 1 = \sum_{i=1}^{r} \left(\hat{b}_{Z_i} \sigma_{Z_i}/\sigma_Y\right)^2, \tag{1.21}$$

so that

$$\sum_{i=1}^{r} \left(\hat{b}_{Z_i} \sigma_{Z_i}\right)^2 = \sigma_Y^2 = V(Y), \tag{1.22}$$

where $V(Y)$ indicates the variance of Y. Equation (1.22) is to highlight that both Equations (1.12) and (1.20) are variance decomposition formulas. As a

---

[15] This discussion holds for linear regression. More sophisticated metamodelling techniques which can overcome these shortcomings are described in Chapter 5.

sensitivity analysis tool, these formulas allow us to decompose the variance of the model output, taken as a descriptor of output uncertainty. Although most practitioners tend to agree on this usage of variance as a proxy for uncertainty, one should remember that the two things are not identical. For example, a measure of uncertainty could be defined on the basis of entropy of model output (see Saltelli *et al.*, 2000, pp. 56–57). In this book we shall use variance decomposition schemes for sensitivity analysis whenever the setting of the analysis allows it.

Wrapping up the results so far, we have seen formulas for decomposing the variance of the model output of interest according to the input factors. Yet we would like to do this for all models, independently of their degree of linearity; that is, we would like to be able to decompose the variance of Y even for models with a low R_Y^2. We want to find what is referred to in the literature as a 'model-free' approach. One such 'model-free' sensitivity measure is based on averaged partial variances, which we now move on to describe along two separate lines.

1.2.6 Conditional Variances – First Path

We have a generic model

$$Y = f(X_1, X_2, \ldots, X_k) \qquad (1.23)$$

like model (1.3) above. Each X has a nonnull range of variation or uncertainty and we wish to determine what would happen to the uncertainty of Y if we could fix a factor. Imagine that we fix factor X_i at a particular value x_i^*. Let $V_{\mathbf{X}_{\sim i}}(Y \mid X_i = x_i^*)$ be the resulting variance of Y, taken over $\mathbf{X}_{\sim i}$ (all factors but X_i). We call this a conditional variance, as it is conditional on X_i being fixed to x_i^*. We would imagine that, having frozen one potential source of variation (X_i), the resulting variance $V_{\mathbf{X}_{\sim i}}(Y \mid X_i = x_i^*)$ will be less than the corresponding total or unconditional variance $V(Y)$. One could therefore conceive of using $V_{\mathbf{X}_{\sim i}}(Y \mid X_i = x_i^*)$ as a measure of the relative importance of X_i, reasoning that the smaller $V_{\mathbf{X}_{\sim i}}(Y \mid X_i = x_i^*)$, the greater the influence of X_i. There are two problems with this approach. First, it makes the sensitivity measure dependent on the position of point x_i^* for each input factor, which is impractical. Second, one can design a model that for particular factors X_i and fixed point x_i^* yields $V_{\mathbf{X}_{\sim i}}(Y \mid X_i = x_i^*) > V(Y)$, i.e. the conditional variance is in fact greater than the unconditional (see the Exercises at the end of this chapter). If we take instead the average of this measure over all possible points x_i^*, the dependence on x_i^* will disappear.

METHODS AND SETTINGS FOR SENSITIVITY ANALYSIS

We write this as $E_{X_i}\left(V_{\mathbf{X}_{\sim i}}(Y \mid X_i)\right)$. This is always lower or equal to $V(Y)$, and in fact:

$$E_{X_i}\left(V_{\mathbf{X}_{\sim i}}(Y \mid X_i)\right) + V_{X_i}\left(E_{\mathbf{X}_{\sim i}}(Y \mid X_i)\right) = V(Y). \tag{1.24}$$

Hence a small $E_{X_i}\left(V_{\mathbf{X}_{\sim i}}(Y \mid X_i)\right)$, or a large $V_{X_i}\left(E_{\mathbf{X}_{\sim i}}(Y \mid X_i)\right)$, will imply that X_i is an important factor. Note that, by Equation (1.24), $V_{X_i}\left(E_{\mathbf{X}_{\sim i}}(Y \mid X_i)\right) \leq V(Y)$. The conditional variance $V_{X_i}\left(E_{\mathbf{X}_{\sim i}}(Y \mid X_i)\right)$ is called the first-order effect of X_i on Y and the sensitivity measure:

$$S_i = \frac{V_{X_i}\left(E_{\mathbf{X}_{\sim i}}(Y \mid X_i)\right)}{V(Y)} \tag{1.25}$$

is known as the first-order sensitivity index of X_i on Y. S_i is a number always between 0 and 1.[16] A high value signals an important variable. And vice versa? Does a small value of S_i flag a nonimportant variable? We leave this question for later and move directly on to the second path for S_i.

1.2.7 Conditional Variances – Second Path

Let us go back to the scatterplots of Figure 1.6. We have said before that what identifies an important factor is the existence of 'shape' or 'pattern' in the points, while a uniform cloud of points is a symptom (though not a proof) of a noninfluential factor. What, then, constitutes shape? We could say that we have a pattern when the distribution of Y-points over the abscissa, i.e. over the factor X_i, is nonuniform. In other words, if the X_i axis is cut into slices, does one see differences in the distribution of Y-points over the slices (Figure 1.7)? Does the mean value of Y in each slice vary across the slices (Figure 1.8)? From Figure 1.7 (which is the same as Figure 1.6, with the addition of 'slices') and Figure 1.8 we can see that factor Z_4 is more influential than Z_1, and that the ordering of factors by importance is $Z_4 > Z_3 > Z_2 > Z_1$, according to how much the mean value of Y varies from one slice to another.

We thus suggest as a sensitivity measure the quantity:

Variation over the slices of the expected value of Y within each slice.

---

[16] Here and in the following we shall tend to use the synthetic notation S_i when the factors considered are labelled X, while we use the lengthier notation, e.g S_{Z_i} or S_{Ω_i}, when the factor has a symbol different from X.

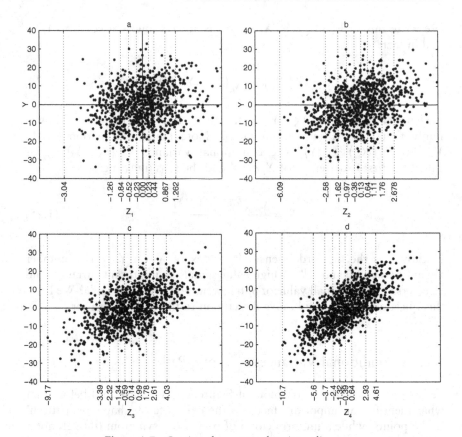

Figure 1.7 Cutting the scatterplots into slices . . .

Taking the limit of this for very thin slices we rediscover $V_{X_i}(E_{\mathbf{X}_{\sim i}}(Y \mid X_i))$. Note indeed that the expected value of Y over a very thin slice corresponds to keeping X_i fixed while averaging over all-but-X_i, which is exactly $E_{\mathbf{X}_{\sim i}}(Y \mid X_i)$. The variance operator is also easily understood.

The issue of cutting the scatterplot into slices will be taken up again in Chapter 5 in the context of metamodelling, at which point a useful approximation of the function expectation value in the slices will be introduced. We anticipate here that $E_{\mathbf{X}_{\sim i}}(Y \mid X_i)$ will be the best predictor of Y based on X_i.

1.2.8 Application to Model (1.3)

Having defined the new sensitivity measure S_i we are eager to apply it to our model of Equation (1.3). It will come as no surprise that for our well-behaved, linear model we obtain

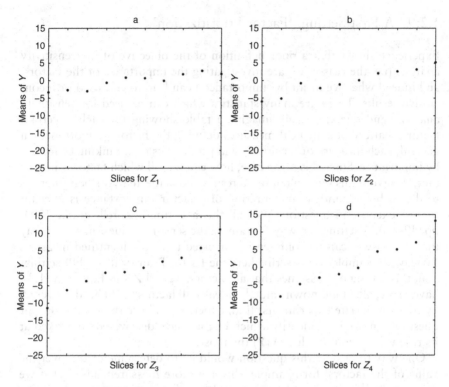

Figure 1.8 ... and taking the average within each slice. Looking at the ordinate, it is clear that Z_3 and Z_4 control more variation than Z_1 and Z_2

$$S_{Z_i} = \frac{V_{Z_i}\left(E_{Z_{\sim i}}(Y \mid Z_i)\right)}{V(Y)} = \beta_{Z_i}^2 \qquad (1.26)$$

(See Table 1.2 for a comparison between $\beta_{Z_i}^2$ and the analytic value of S_{Z_i}.) The identity of Equation (1.26) holds for linear models, as we would expect given that S_{Z_i} is a model-free generalization of $\beta_{Z_i}^2$. For nonlinear models the two measures will differ, as we shall see in a moment. Another important difference between S_{Z_i} and $\beta_{Z_i}^2$ is that while $\sum_{i=1}^{r} \beta_{Z_i}^2 = 1$ only for linear models, the relationship $\sum_{i=1}^{r} S_{Z_i} = 1$ holds for a larger class: that of additive models. By definition, a model is additive when it is possible to separate the effects of its input variables in a variance decomposition framework. For example, $Y = \sum_i Z_i^2$ is a nonlinear, additive model in the Z's; $Y = \prod_i Z_i$ is nonlinear and nonadditive.

For nonadditive models the first-order terms do not add up to one, i.e. $\sum_{i=1}^{r} S_{Z_i} \leq 1$. This is also how nonadditive models are defined. We shall turn to this presently, after first discussing the need for 'settings' in sensitivity analysis.

1.2.9 A First Setting: 'Factor Prioritization'

Experience shows that a poor definition of the objective of the sensitivity analysis (i.e. the reason we are investigating the importance of the factors, and indeed what we mean by 'importance') can lead to confused or inconclusive results. There are many statistics which can be used for sensitivity analysis, and one can easily imagine a table showing for each uncertain factor a battery of statistical measures defining the factor's importance. In general, each measure of sensitivity will produce its own ranking of factors by importance. Since this is the case, how can we tell which factor is important? To avoid this kind of confused result, it is in the analyst's best interests to define beforehand what definition of a factor's importance is relevant for the exercise in question. We call this a 'setting' (Saltelli et al., 2004, pp. 49–56). A setting is a way of framing the sensitivity quest in such a way that the answer can be confidently entrusted to a well-identified measure. By way of example, we describe here the Factor Prioritization (FP) setting.

In this setting one assumes that all factors, e.g. all Z's in Equation (1.3), have a true, albeit unknown value.[17] Ideally all factors could be 'discovered' by the appropriate experiments. If all experiments have the same cost, our quest or venture is to identify which factor, once 'discovered' and fixed at its true value, would reduce $V(Y)$ the most.

One way to answer this question would be to determine or discover the value of the factors, for example, through more measurements. Yet if we were to do this we would have gone beyond uncertainty and sensitivity analyses. The challenge, therefore, is to identify the appropriate factors before any of them are measured or discovered, i.e. when the value to which each factor should be fixed is unknown. This suggests that a good contender for the title of 'most influential factor' would be that factor which, on average, once fixed, would cause the greatest reduction in variance. 'On average', in this case, means that we must average the fixing of the factor over the distribution of the factor itself. It is straightforward to see that in this setting $E_{X_i}(V_{\mathbf{X}_{\sim i}}(Y \mid X_i))$ is the measure to use. The lower $E_{X_i}(V_{\mathbf{X}_{\sim i}}(Y \mid X_i))$, and hence the higher $V_{X_i}(E_{\mathbf{X}_{\sim i}}(Y \mid X_i))$, the more probable it is that factor X_i is the factor that one should measure first in order to reduce the variance most. We have thus linked the FP setting to a measure, the first-order sensitivity index S_i. This is a gamble, as we do not know

---

[17] In most circumstances one will have factors susceptible of determination, for which a true unknown value can be hypothesized (e.g. the failure rate of a component type, the value of an activation energy for a chemical reaction), as well as factors intrinsically uncertain (the time of failure of a specific component, the wind direction at a given time and location). These are termed epistemic and stochastic uncertainties respectively. For the purpose of illustrating the setting it is convenient to imagine all factors epistemically uncertain.

the position of the true value of a factor over its distribution. Someone actually measuring a given factor could still beat our sensitivity analysis-based guess and reduce the variance by more than we have guessed, or reduce the variance using a factor other than the one we identified via sensitivity analysis.

1.2.10 Nonadditive Models

In order to gain confidence with nonadditivity in models, we return to the input for our elementary model (1.3), $Y = \sum_{i=1}^{r} \Omega_i Z_i$, and complicate it by allowing both the Z's and the Ω's to become factors – the Ω's are no longer constants. We do this to generate a nonadditive model, as we shall see presently. The additivity of a model depends upon the characteristics of its input factors, so that it is sufficient to change a constant of the model into a factor in order to change the model from additive to nonadditive, although the model is left unchanged in the form (1.3). Our input description becomes

$$\begin{array}{ll} Z_i \sim N(\bar{z}_i, \sigma_{Z_i}) & \bar{z}_i = 0 \\ \Omega_i \sim N(\bar{\Omega}_i, \sigma_{\Omega_i}) & \bar{\Omega}_i = ic \end{array} \quad i = 1, 2, \ldots, r. \quad (1.27)$$

The distribution of the Z's remains unchanged, while the Ω's, so far constant, become input factors with normal distribution. Their mean is not zero as it is for the Z's, but rather some number other than zero – we shall explain why in the Exercises at the end of this chapter. For the sake of the example we have made the means of the Ω's nonequivalent and equal in value to the product of the integer i (used as counter) and a positive constant c. This is simply a way to make the means of the Ω's increase, so that Equation (1.7) is no longer true. Instead

$$\bar{\Omega}_1 < \bar{\Omega}_2 < \ldots \bar{\Omega}_r. \quad (1.28)$$

The input factors for the analysis are

$$\mathbf{X} = (Z_1, Z_2, \ldots, Z_r, \Omega_1, \Omega_2, \ldots, \Omega_r) \quad (1.29)$$

and the total number of factors is $k = 2r$. We now perform another Monte Carlo experiment, sampling both the Z's and the Ω's from their respective distributions in Equation (1.27). Remember that we assume all factors independent, so each factor is sampled from its marginal distribution with no consideration of where the other factors are sampled. How the Monte Carlo sample is used to produce estimates $\hat{S}_i$ of the first-order sensitivity measures S_i is explained later in this book (see Chapter 4). We anticipate the results

Table 1.3 First-order indices S_i (analytic) and squared standardized regression coefficient β_i^2 for model (1.3, 1.27), where $r = 4$, $c = 0.5$, $\sigma = (1, 2, 3, 4)$ for both Ω_i and Z_i, and $N = 40.000$ for the regression analysis. Such a large sample was used to show the convergence between S_i and β_i^2

	S_i	β_i^2
Z_1	0.0006	0
Z_2	0.009	0.01
Z_3	0.046	0.05
Z_4	0.145	0.14
Ω_1	0	0
Ω_2	0	0
Ω_3	0	0
Ω_4	0	0

in Table 1.3, where the squared standardized regression estimates $\hat{\beta}^2$ are also reported for comparison.

It is evident from Table 1.3 that while $\hat{S}_{Z_i}$ are still greater than zero, the $\hat{S}_{\Omega_i}$ are practically zero. Furthermore

$$\sum_{i=1}^{k} \hat{S}_{X_i} = \sum_{i=1}^{r} \hat{S}_{Z_i} + \sum_{i=1}^{r} \hat{S}_{\Omega_i} < 1. \qquad (1.30)$$

We had already anticipated that for a nonadditive model the sum of the first-order indices would be less than one.

However, it might seem puzzling that the Ω input factors seem to have no influence. In fact, it is not difficult to understand why S_{Ω_i} must be zero (Figure 1.9).

Let us go back to our definition of S_i, Equation (1.25):

$$S_i = \frac{V_{X_i}\left(E_{\mathbf{X}_{\sim i}}(Y \mid X_i)\right)}{V(Y)}, \qquad (1.31)$$

and let us compute it for Ω_i,

$$S_{\Omega_i} = \frac{V_{\Omega_i}\left(E_{\mathbf{X}_{\sim \Omega_i}}(Y \mid \Omega_i)\right)}{V(Y)}. \qquad (1.32)$$

We focus on the inner expectation $E_{\mathbf{X}_{\sim \Omega_i}}(Y|\Omega_i)$ which we now have to write explicitly as $E_{\mathbf{X}_{\sim \Omega_i}}(Y \mid \Omega_i = \omega_i^*)$ in order to remind ourselves that we have fixed Ω_i.

METHODS AND SETTINGS FOR SENSITIVITY ANALYSIS

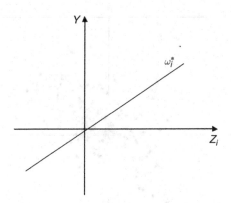

Figure 1.9 Y versus Z_i for fixed values of $X_{\sim Z_i}$

Note that $E_{X_{\sim \Omega_i}}$ now means that the mean is taken over all Z_j's, including Z_i, and over all Ω_j's but Ω_i.

Figure 1.9 shows the plot of Y versus Z_i for a fixed nonzero value of ω_i^* of input factor Ω_i in the case that all the remaining Ω_j's, with $j \neq i$, are fixed to zero. This straight line will be shifted up or down vertically when the Ω_j's, with $j \neq i$, are fixed to values other than zero.

Positive and negative values of Y will hence be equally probable and equally distributed, so that $E_{X_{\sim \Omega_i}}(Y \mid \Omega_i = \omega_i^*)$ will be zero. Figure 1.10 shows how this emerges from Monte Carlo generated scatterplots of Y versus Z_i and Y versus Ω_i. It is clear that if $E_{X_{\sim \Omega_i}}(Y \mid \Omega_i = \omega_i^*)$ is zero for any value ω_i^*, its variance over all possible values of ω_i^* will also be zero, so that both $V_{\Omega_i}(E_{X_{\sim \Omega_i}}(Y \mid \Omega_i))$ and S_{Ω_i} will be zero for all factors Ω_i.

We now understand that the measure S_{Ω_i} is zero, but we retain the belief that factors Ω_i should have some influence, especially since this is suggested by the conical pattern evident in Figure 1.10. It seems therefore that there may be a problem with our sensitivity measure. A regression coefficient $\hat{\beta}_{\Omega_i}$ would produce a straight horizontal line through the horizontal conical plot of Y versus Ω_i in Figure 1.10. However, it is clear from the shape of this plot that variable Ω_i is influential. A possible interpretation is that $\hat{\beta}_{\Omega_i}$ fails as a sensitivity measure in this case. Does the fact that S_{Ω_i} is zero imply that also S_{Ω_i} fails?

Indeed it is unfair to say that β_{Ω_i} fails in Figure 1.10. β_{Ω_i} is a linear measure, so clearly it should not be used on a nonlinear model. S_{Ω_i}, however, is a model-free measure, and must be applicable to nonlinear models. Indeed this is the case, and we can say that if S_{Ω_i} is zero, this means that Ω_i has no effect on Y 'at the first order' (recall that we have thus far discussed first-order sensitivity indices). The reader familiar with experimental design

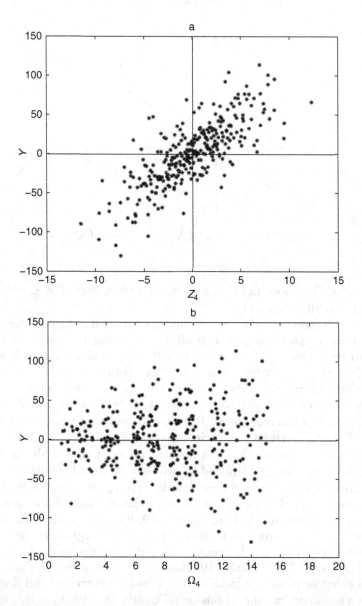

Figure 1.10 Scatterplots of Y versus Z_4 and versus Ω_4 for model (1.3, 1.27), at sample size $N = 1000$. The first-order sensitivity index for Z_4 is greater than zero while that for Ω_4 is zero

METHODS AND SETTINGS FOR SENSITIVITY ANALYSIS

will already have guessed that the effect of Ω_i must be captured by some higher-order effect, as we now proceed to discuss.

1.2.11 Higher-order Sensitivity Indices

We continue our game with conditioned variances by playing with two factors instead of one. Take for instance

$$\frac{V(E(Y \mid Z_i, Z_j))}{V(Y)}, \qquad (1.33)$$

with $i \neq j$. We have dropped the indices of both the E and V operators. Indeed we do not need them if we accept the convention that the argument conditioning the inner operator, Z_i, Z_j in this case, is also the set over which we apply the outer operator, i.e. the variance is taken over Z_i, Z_j (we should have written V_{Z_i,Z_j}). By default, the inner operator, the average E, must be taken over all-but-(Z_i, Z_j). What would happen if we could compute (1.33), with $i \neq j$, and compare it with the corresponding measure for the individual factors Z_i, Z_j? We would observe that

$$\frac{V(E(Y \mid Z_i, Z_j))}{V(Y)} = S_{Z_i} + S_{Z_j} \quad \text{for} \quad i \neq j, \qquad (1.34)$$

while

$$\frac{V(E(Y \mid \Omega_i, \Omega_j))}{V(Y)} = 0, \qquad (1.35)$$

and

$$\frac{V(E(Y \mid Z_i, \Omega_i))}{V(Y)} > S_{Z_i} + S_{\Omega_i}. \qquad (1.36)$$

We anticipate from Chapter 4 that, given two generic factors X_i, X_j, the following result holds:

$$V(E(Y \mid X_i, X_j)) = V_i + V_j + V_{ij}, \qquad (1.37)$$

where

$$\begin{aligned} V_i &= V(E(Y \mid X_i)) \\ V_j &= V(E(Y \mid X_j)) \\ V_{ij} &= V(E(Y \mid X_i, X_j)) - V_i - V_j. \end{aligned} \qquad (1.38)$$

The term V_{ij} is the interaction term between factors X_i, X_j. It captures that part of the response of Y to X_i, X_j that cannot be written as a superposition of effects separately due to X_i and X_j. Recalling our previous examples of $Y = \sum_i Z_i^2$ (a nonlinear, additive model) and $Y = \prod_i Z_i$ (nonlinear, nonadditive), the latter model will have nonzero second-order terms such as V_{ij}, while the former model will not.

Looking at Equations (1.37, 1.39) and remembering that for our model all S_{Ω_i} are zero, we are now ready to grasp the results of Equations (1.34–1.36) (see also Table 1.4).

- Equation (1.34) holds because the interaction term between Z_i and Z_j is zero, which is evident from the form of Equation (1.3).
- Equation (1.35) holds because the S_{Ω_i} and S_{Ω_j} as well as their interaction term are zero.
- Equation (1.36) can be rewritten as

$$\frac{V(E(Y \mid Z_i, \Omega_j))}{V(Y)} = S_{Z_i} + S_{\Omega_i} + S_{Z_i, \Omega_i},$$

where $S_{\Omega_i} = 0$, $S_{Z_i, \Omega_i} = V_{Z_i, \Omega_i}/V(Y)$ and the term V_{Z_i, Ω_i} is the only type of nonzero second-order term in model (1.3).

If we now sum all the nonzero first-order and and second-order terms we get

Table 1.4 First- and second-order indices for model (1.3, 1.27, analytic), where $r = 4$, $c = 0.5$, $\sigma = (1, 2, 3, 4)$ for both Ω_i and Z_i

Factor	S_i, S_{ij}	Factor	S_{ij}	Factor	S_{ij}
Z_1	0.0006	Z_1, Ω_2	0	Z_3, Ω_3	0.183
Z_2	0.009	Z_1, Ω_3	0	Z_3, Ω_4	0
Z_3	0.046	Z_1, Ω_4	0	Z_4, Ω_1	0
Z_4	0.145	Z_2, Z_3	0	Z_4, Ω_2	0
Ω_1	0	Z_2, Z_4	0	Z_4, Ω_3	0
Ω_2	0	Z_2, Ω_1	0	Z_4, Ω_4	0.578
Ω_3	0	Z_2, Ω_2	0.036	Ω_1, Ω_2	0
Ω_4	0	Z_2, Ω_3	0	Ω_1, Ω_3	0
Z_1, Z_2	0	Z_2, Ω_4	0	Ω_1, Ω_4	0
Z_1, Z_3	0	Z_3, Z_4	0	Ω_2, Ω_3	0
Z_1, Z_4	0	Z_3, Ω_1	0	Ω_2, Ω_4	0
Z_1, Ω_1	0.002	Z_3, Ω_2	0	Ω_3, Ω_4	0

METHODS AND SETTINGS FOR SENSITIVITY ANALYSIS

$$\sum_{t=1}^{r} (S_{Z_i} + S_{Z_i \Omega_i}) = 1. \tag{1.39}$$

This means that even for a nonadditive model we have found a way to recover (that is, to understand) 100% of the variance of Y. Thus variance-based sensitivity measures provide a theoretical framework whereby – provided one has the patience to compute all interaction terms – one can achieve a full understanding of the model's sensitivity pattern. Patience is indeed required, as in principle a model can have interactions of even higher order. Again anticipating one result from Chapter 4, a full analysis of a model with k factors is composed of

$$\sum_i S_i + \sum_i \sum_{j>i} S_{ij} + \sum_i \sum_{j>i} \sum_{l>j} S_{ijl} + \ldots + S_{123\ldots k} = 1. \tag{1.40}$$

Model (1.3) can only have nonzero terms up to the second order, and this can be seen 'by inspection', as the structure of the model is very simple. In practical applications the subject model of our analysis will be a computer program, and the only way to ascertain whether an interaction exists or not will be to estimate it numerically. The problem is that the series development of Equation (1.40) has as many as $2^k - 1$ terms. For $k = 3$ this gives just 7 terms, i.e. $S_1, S_2, S_3, S_{12}, S_{23}, S_{13}, S_{123}$; for $k = 10$ it gives 1023, too many to look at in practice.

In fact, the variance-based analysis can help us in these circumstances, by computing for each factor a 'total effect' term, which we describe next.

1.2.12 Total Effects

What is a total effect term? Let us again use our extended model (1.3, 1.27), and ask what we would obtain if we were to compute $V(E(Y | \mathbf{X}_{\sim \Omega_i}))/V(Y)$. We are conditioning now on all factors but Ω_i. In other words

$$\frac{V(E(Y|\mathbf{X}_{\sim \Omega_i}))}{V(Y)} = \frac{V(E(Y | \Omega_1, \Omega_2, \ldots, \Omega_{i-1}, \Omega_{i+1}, \ldots \Omega_r, Z_1, Z_2, \ldots, Z_r))}{V(Y)}. \tag{1.41}$$

By analogy with our discussion of second-order terms, Equation (1.41) should include all terms of any order that do not include factor Ω_i. As the sum of all possible sensitivity terms must be 1, the difference

$$\left(1 - \frac{V(E(Y|\mathbf{X}_{\sim \Omega_i}))}{V(Y)}\right)$$

must be made up of all terms of any order that include ω_i. For our model, which has only first- and second-order terms, this gives

$$\left(1 - \frac{V(E(Y|\mathbf{X}_{\sim \Omega_i}))}{V(Y)}\right) = S_{\Omega_i} + S_{Z_i \Omega_i} \quad \text{(See Table 1.5)} \quad (1.42)$$

To consider a different example, for a generic three-factor model, one would have

$$S_{T1} = \left(1 - \frac{V(E(Y|\mathbf{X}_{-1}))}{V(Y)}\right) = S_1 + S_{12} + S_{13} + S_{123} \quad (1.43)$$

and

$$S_{T2} = S_2 + S_{12} + S_{23} + S_{123}$$
$$S_{T3} = S_3 + S_{13} + S_{23} + S_{123},$$

where S_{Ti} denotes the total effect of factor X_i. We recall that we tend to use the synthetic notation $(S_i, S_{Ti}, V_i, S_{ij})$ when the factors considered are labelled X, while we use the lengthier notation $(S_{Z_i}, S_{TZ_i}, V_{\Omega_i}, S_{Z_i \Omega_i})$ when the factor has a symbol other than X.

Table 1.5 First-order and total effects for model (1.3, 1.27, analytic), where $r = 4, c = 0.5, \sigma = (1, 2, 3, 4)$ for both Ω_i and Z_i

	S_i		S_{Ti}
Z_1	0.0006	Z_1	0.003
Z_2	0.009	Z_2	0.045
Z_3	0.046	Z_3	0.229
Z_4	0.145	Z_4	0.723
Ω_1	0	Ω_1	0.002
Ω_2	0	Ω_2	0.036
Ω_3	0	Ω_3	0.183
Ω_4	0	Ω_4	0.578

We have argued in a series of works (Saltelli *et al.*, 2004, and references therein) that a good, synthetic, though nonexhaustive characterization of the sensitivity pattern for a model with k factors is given by the total set of first-order terms plus the total effects. For a system with 10 factors this makes 20 terms rather than 1023.

One last observation about the total effect terms is the following. For the algebraic rule already mentioned in Equation (1.24) we have

$$E_{X_i}\left(V_{\mathbf{X}_{\sim i}}(Y \mid X_i)\right) + V_{X_i}\left(E_{\mathbf{X}_{\sim i}}(Y \mid X_i)\right) = V(Y),$$

and hence

$$S_{Ti} = 1 - \frac{V(E(Y \mid \mathbf{X}_{\sim i}))}{V(Y)} = \frac{E(V(Y \mid \mathbf{X}_{\sim i}))}{V(Y)}. \quad (1.44)$$

Equipped with this new sensitivity measure, the total effect, we are now ready to introduce another useful 'setting' for sensitivity analysis.

1.2.13 A Second Setting: 'Factor Fixing'

One use of sensitivity analysis is to simplify models. If a model is used systematically in a Monte Carlo framework, so that input uncertainties are always propagated through the output, it might be useful to ascertain which of the input factors can be fixed anywhere in their range of variation without appreciably affecting a specific output of interest Y. This could help to simplify a model in a greater sense, since we might be able to condense (lump) an entire section of our model if all factors entering that section are noninfluential. From the preceding discussion it will be clear that $S_i = 0$ is a necessary but insufficient condition for fixing factor X_i. This factor might be involved in interactions with other factors such that, although its first-order term is zero, there might be nonzero higher-order terms. This is exactly what happened with our factors Ω_i in the model (1.3, 1.27).

Imagine now that a factor X_i is truly noninfluential. Let us compute $V_{X_i}(Y \mid \mathbf{X}_{\sim i} = \mathbf{x}^*_{\sim i})$, where we have fixed a point $\mathbf{x}^*_{\sim i}$ in the multidimensional space $\mathbf{X}_{\sim i}$. If factor X_i is noninfluential, then $V_{X_i}(Y \mid \mathbf{X}_{\sim i} = \mathbf{x}^*_{\sim i})$ must be zero, as the value of Y is totally determined by $\mathbf{X}_{\sim i}$ and there will be no variance over X_i. Averaging over non-X_i will not change the result, so that $E_{\mathbf{X}_{\sim i}}\left(V_{X_i}(Y \mid \mathbf{X}_{\sim i})\right)$ must be zero as well. Based on our convention of not indicating the conditioning argument, we can also write this as $E(V(Y \mid \mathbf{X}_{\sim i})) = 0$. These considerations prove that if X_i is noninfluential, then $S_{Ti} = 0$ by Equation (1.44) above.

Conversely, if $S_{Ti} = 0$ for factor X_i, then $E(V(Y \mid \mathbf{X}_{\sim i})) = 0$. As the variance can only be a positive number, the fact that the mean $E(V(Y \mid \mathbf{X}_{\sim i}))$ is zero implies that $V(Y \mid \mathbf{X}_{\sim i} = \mathbf{x}^*_{\sim i})$ is identically zero for any value of $\mathbf{x}^*_{\sim i}$, which proves that X_i is noninfluential – there is no point in the hyperspace of $\mathbf{X}$ where X_i has an effect. This demonstrates that $S_{Ti} = 0$ is a necessary and sufficient condition for X_i being noninfluential.

Note that the model simplification underpinned by the 'factor fixing' setting can become very important when models need to be audited, for example in the face of a scientific controversy or for use in policy assessment. In these situations one might wish to optimize the 'relevance' R of a model, defined as the ratio (Beck et al., 1997):

$$R = \frac{\text{number of factors that truly induce variations in the output of interest}}{\text{total number of factors in the model}}.$$

This approach would guard against the criticism that an overly complex model was being used by one party to obfuscate or discourage investigation.

The concepts of parsimony or simplicity in the context of modelling are illustrated by the works of Peter C. Young (Young et al., 1996; Young, 1999a), who recommends the use of data-driven models, in which a minimum of parameters are inferred directly from the data, as an alternative to law-driven, usually overparametrized models. To give an example, the hydrogeology of a catchment area can be modelled with a complex model based on Darcy's laws or with a low-order model based on direct interpretation of precipitation and runoff time series. Such a parsimonious description of the system can also be thought of as a complement to a law-driven model. More generally, for models to be used in impact assessment or other regulatory settings, it might be advisable to have a back-of-the-envelope version of the general model for the purpose of negotiating assumptions and inferences with stakeholders. Sensitivity analysis may be instrumental in deriving such a simplified model.

The foregoing discussion of possible settings for sensitivity analysis allows us to make a few more observations on the rationale for sensitivity analysis.

1.2.14 Rationale for Sensitivity Analysis

Possible motivations for sensitivity analysis are:

- Model corroboration. Is the inference robust? Is the model overly dependent on fragile assumptions?

- Research prioritization. Which factor is most deserving of further analysis or measurement? ⇒ factor prioritization setting.
- Model simplification. Can some factors or compartments of the model be fixed or simplified? ⇒ factor fixing setting.
- Identifying critical or otherwise interesting regions in the space of the input factors. Identifying factors which interact and which may thus generate extreme values. This is important, for example in system reliability.
- Prior to parameter estimation, to help set up the (actual or numerical) experiment in those conditions in which the sensitivity of the output to the factor to be estimated is the greatest.

To illustrate the last point, imagine that one has actual measurements against which to compare model predictions. Ideally, predictions and measurements can feed into an estimation step. Yet before this is done, it is worth investigating what drives, for instance, the sum of the squared differences between model prediction and actual measurements. Only factors with this type of influence are good candidates for the estimation step. In this way the analyst can decide which experimental conditions are more interesting for the subsequent estimation (Saltelli *et al.*, 2004, pp.151–191).

We have already mentioned that uncertainty and sensitivity analyses can be run in tandem to ascertain whether different policies (e.g. strategies to alleviate an environmental problem) are indeed different from one another when compared in the overall space of the uncertainties. An example of such an analysis is found in Saltelli *et al.* (2000 pp. 385–397).

It is worth noting in this case that high uncertainty in the inference is not synonymous with low quality in the resulting assessment. Though uncertain, the assessment might still allow policy A to be distinguished from policy B (implying high quality) while the opposite is also possible, i.e. that the model might not allow these options to be distinguished even with only moderate uncertainties in the inference (implying a low-quality assessment). On a similar ground, when confronted with a plurality of stakeholders' views and beliefs as to how an issue should be tackled or framed, we may use sensitivity analysis to ascertain whether – within the latitude of the different framings and assumptions – we still can reach some robust inference, i.e. a high-quality assessment. We would call such an inference – or the resulting preferred policy – 'socially' robust, as it is compatible with such a plurality of viewpoints. On the contrary, we might find that the different framings give rise to such great latitude in the resulting inference that no robust policy can be identified.

Another general consideration with respect to the global, explorative nonparametric methods for the sensitivity analysis just described is that these have a better chance of being resilient towards type II errors than local (derivative-based) methods. The possibility of important factors being overlooked or dangerous or critical combinations of input factors neglected decreases with the level of exploration of the space of the input factors (Farrell, 2007). The attention paid in global methods to interaction effects is also a protection against type II errors. In Saltelli *et al.* (2005) we show that, for even a relatively simple and well-studied chemical reactor system, global sensitivity analysis and attention to the interactions can lead to the identification of a larger 'runaway' portion in the space of the input factors than could previously be identified.

Some of the motivations just described would demand being able to apportion uncertainty not only among factors, but also among sets of factors, for example to distinguish data uncertainty from experts' uncertainty, system uncertainty from policy option uncertainty and so on. We offer a few tools for this in the following.

1.2.15 Treating Sets

An additional interesting feature of variance-based methods is that they allow for a concise treatment of the sensitivity of sets of factors. Referring again to model (1.3, 1.27), we can imagine computing a variance measure conditioned on a subset of the input factors, e.g. on the set Ω, $S_\Omega = V(E(Y \mid \Omega))/V(Y)$. From the description in the previous sections it is easy to understand that S_Ω will include all first-order terms related to Ω plus second- and higher-order product terms including only members of Ω. We already know that these are all zero. We can likewise compute $S_Z = V(E(Y \mid Z))/V(Y)$ for the set Z. This similarly contains all nonzero first-order terms plus the null second- and higher-order terms internal to Z. Finally we can compute

$$S_{\Omega,Z} = V(Y) - S_\Omega - S_Z, \qquad (1.45)$$

which will contain all cross-product terms not involved in S_Ω, S_Z. Going back to our example of Equation (1.3) as a composite indicator with weights Ω given by experts and variables Z coming from statistical offices, with Equation (1.45) we have apportioned variance between data and experts and an interaction between the two.

Similarly, we could share the uncertainty in Y among the couples $A_i = (\Omega_i, Z_i)$ and apply

$$\sum_i S_{A_i} + \sum_i \sum_{j>i} S_{A_i A_j} + \sum_i \sum_{j>i} \sum_{l>j} S_{A_i A_j A_l} + \ldots = 1. \qquad (1.46)$$

METHODS AND SETTINGS FOR SENSITIVITY ANALYSIS

As we already know that for our model all cross-product terms with $i \neq j$ are zero, this can be reduced to the convenient

$$\sum_i^r S_{A_i} = 1, \quad (1.47)$$

in which uncertainty is divided among 'themes', each theme comprising an indicator and its weight. It is easy to imagine similar applications. For example, one could divide uncertainty among observational data, estimation, model assumptions, model resolution and so on.

1.2.16 Further Methods

So far we have discussed the following tools for sensitivity analysis:

- derivatives and sigma-normalized derivatives;
- regression coefficients (standardized);
- variance-based measures;
- scatterplots.

We have shown the equivalence of sigma-normalized coefficients $S_i^\sigma = \sigma_{Z_i} \partial Y / \sigma_Y \partial X_i$, regression coefficients β_i and variance-based first-order sensitivity indices S_i for linear models, as well as how S_i is a model-free extension of the variance decomposition scheme to models of unknown linearity. We have discussed how nonadditive models can be treated in the variance-based sensitivity framework. We have also indicated that scatterplots are a powerful tool for sensitivity analysis and shown how S_i can be interpreted in relation to the existence of 'shape' in an X_i versus Y scatterplot.

At a greater level of detail (Ratto *et al.*, 2007) one can use modern regression tools (such as state-space filtering methods) to interpolate points in the scatterplots, producing very reliable $E(Y \mid X_i = x_i^*)$ curves. The curves can then be used for sensitivity analysis. Their shape is more evident than that of dense scatterplots (compare Figure 1.7 with Figure 1.8). Furthermore, one can derive the first-order sensitivity indices directly from those curves, so that an efficient way to estimate S_i is to use state-space regression on the scatterplots and then take the variances of these.

In general, for a model of unknown linearity, monotonicity and additivity, variance-based measures constitute a good means of tackling settings such as factor fixing and factor prioritization. We shall discuss one further setting before the end of this chapter, but let us first consider whether there are alternatives to the use of variance-based methods for the settings so far described.

Why might we need an alternative? The main problem with variance-based measures is computational cost. Estimating the sensitivity coefficients takes many model runs (see Chapter 4). Accelerating the computation of sensitivity indices of all orders – or even simply of the S_i, S_{Ti} couple – is the most intensely researched topic in sensitivity analysis (see the filtering approach just mentioned). It can reasonably be expected that the estimation of these measures will become more efficient over time.

At the same time, and if only for screening purposes, it would be useful to have methods to find approximate sensitivity information at lower sample sizes. One such method is the Elementary Effect Test.

1.2.17 Elementary Effect Test

The Elementary Effect Test is simply an average of derivatives over the space of factors. The method is very simple. Consider a model with k independent input factors $X_i, i = 1, 2, \ldots, k$, which varies across p levels. The input space is the discretized p-level grid Ω. For a given value of $\mathbf{X}$, the elementary effect of the ith input factor is defined as

$$EE_i = \frac{[Y(X_1, X_2, \ldots, X_{i-1}, X_i + \Delta, \ldots X_k) - Y(X_1, X_2, \ldots, X_k)]}{\Delta}, \quad (1.48)$$

where p is the number of levels, Δ is a value in $\{1/(p-1), \ldots, 1-1/(p-1)\}$, $X = (X_1, X_2, \ldots X_k)$ is any selected value in Ω such that the transformed point $(\mathbf{X} + \mathbf{e}_i \Delta)$ is still in Ω for each index $i = 1, \ldots, k$, and $\mathbf{e}_i$ is a vector of zeros but with a unit as its ith component. Then the absolute values of the EE_i, computed at r different grid points for each factor, are averaged

$$\mu_i^* = \frac{1}{r} \sum_{j=1}^{r} |EE_i^j| \quad (1.49)$$

and the factors ranked according to the obtained mean μ_i^*.

In order to compute efficiently, a well-chosen strategy is needed for moving from one effect to the next, so that the input space is explored with a minimum of points (see Chapter 3).

Leaving aside computational issues for the moment, μ^* is a useful measure for the following reasons:

1. It is semi-quantitative – the factors are ranked on an interval scale;
2. It is numerically efficient;
3. It is very good for factor fixing – it is indeed a good proxy for S_{Ti};
4. It can be applied to sets of factors.

Due to its semi-quantitative nature the μ^* can be considered a screening method, especially useful for investigating models with many (from a few dozen to 100) uncertain factors. It can also be used before applying a variance-based measure to prune the number of factors to be considered. As far as point (3) above is concerned, μ^* is rather resilient against type II errors, i.e. if a factor is deemed noninfluential by μ^* it is unlikely to be identified as influential by another measure.

1.2.18 Monte Carlo Filtering

While μ^* is a method of tackling factor fixing at lower sample size, the next method we present is linked to an altogether different setting for sensitivity analysis. We call this 'factor mapping' and it relates to situations in which we are especially concerned with a particular portion of the distribution of output Y. For example, we are often interested in Y being above or below a given threshold. If Y were a dose of contaminant, we might be interested in how much (how often) a threshold level for this contaminant is being exceeded. Or Y could be a loss (e.g. financial) and we might be interested in how often a maximum admissible loss is being exceeded. In these settings we tend to divide the realization of Y into 'good' and 'bad'. This leads to Monte Carlo filtering (MCF, see Saltelli *et al.*, 2004, pp. 151–191 for a review). In MCF one runs a Monte Carlo experiment producing realizations of the output of interest corresponding to different sampled points in the input factor space, as for variance-based or regression analysis. Having done this, one 'filters' the realizations, e.g. elements of the Y-vector. This may entail comparing them with some sort of evidence or for plausibility (e.g. one may have good reason to reject all negative values of Y). Or one might simply compare Y against thresholds, as just mentioned. This will divide the vector Y into two subsets: that of the well-behaved realizations and that of the 'misbehaving' ones. The same will apply to the (marginal) distributions of each of the input factors. Note that in this context one is not interested in the variance of Y as much as in that part of the distribution of Y that matters – for example, the lower-end tail of the distribution may be irrelevant compared to the upper-end tail or vice versa, depending on the problem. Thus the analysis is not concerned with which factor drives the variance of Y as much as with which factor produces realizations of Y in the forbidden zone. Clearly, if a factor has been judged noninfluential by either μ^* or S_{Ti}, it will be unlikely to show up in an MCF. Steps for MCF are as follows:

- A simulation is classified as either B, for behavioural, or $\overline{B}$, for nonbehavioural (Figure 1.11).
- Thus a set of binary elements is defined, allowing for the identification of two subsets for each X_i: one containing a number n of elements denoted

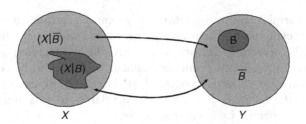

Figure 1.11 Mapping behavioural and nonbehavioural realizations with Monte Carlo filtering

$(X \mid B)$ and a complementary set $\left(X \mid \overline{B}\right)$ containing the remaining $\overline{n} = N - n$ simulations (Figure 1.11).
- A statistical test can be performed for each factor independently, analysing the maximum distance between the cumulative distributions of the $(X \mid B)$ and $\left(X \mid \overline{B}\right)$ sets (Figure 1.12).

If the two sets are visually and statistically[18] different, then X_i is an influential factor in the factor mapping setting.

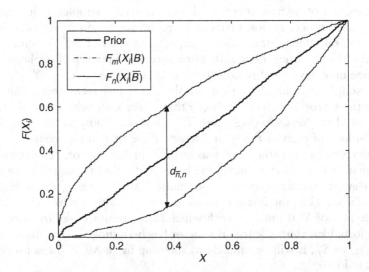

Figure 1.12 Distinguishing between the two sets using a test statistic

---

[18] Smirnov two-sample test (two-sided version) is used in Figure 1.12 (see Saltelli *et al.*, 2004, pp. 38–39).

1.3 NONINDEPENDENT INPUT FACTORS

Throughout this introductory chapter we have systematically assumed that input factors are independent of one another. The main reason for this assumption is of a very practical nature: dependent input samples are more laborious to generate (although methods are available for this; see Saltelli *et al.*, 2000) and, even worse, the sample size needed to compute sensitivity measures for nonindependent samples is much higher than in the case of uncorrelated samples.[19] For this reason we advise the analyst to work on uncorrelated samples as much as possible, e.g. by treating dependencies as explicit relationships with a noise term.[20] Note that when working with the MCF just described a dependency structure is generated by the filtering itself. The filtered factors will probably correlate with one another even if they were independent in the original unfiltered sample. This could be a useful strategy to circumvent the use of correlated samples in sensitivity analysis. Still there might be very particular instances where the use of correlated factors is unavoidable. A case could occur with the parametric bootstrap approach described in Figure 1.3. After the estimation step the factors will in general be correlated with one another, and if a sample is to be drawn from these, it will have to respect the correlation structure.

Another special instance when one has to take factors' dependence into consideration is when analyst A tries to demonstrate the falsity of an uncertainty analysis produced by analyst B. In such an adversarial context, A needs to show that B's analysis is wrong (e.g. nonconservative) even when taking due consideration of the covariance of the input factors as explicitly or implicitly framed by B.

1.4 POSSIBLE PITFALLS FOR A SENSITIVITY ANALYSIS

As mentioned when discussing the need for settings, a sensitivity analysis can fail if its underlying purpose is left undefined; diverse statistical tests and measures may be thrown at a problem, producing a range of different factor rankings but leaving the researcher none the wiser as to which

---

[19] Dependence and correlation are not synonymous. Correlation implies dependence, while the opposite is not true. Dependencies are nevertheless described via correlations for practical numerical computations.
[20] Instead of entering X_1 and X_2 as correlated factors one can enter X_1 and X_3, with X_3 being a factor describing noise, and model X_2 as a function of X_1 and X_3.

ranking to believe or privilege. Another potential danger is to present sensitivity measures for too many output variables Y. Although exploring the sensitivity of several model outputs is sound practice for testing the quality of the model, it is better, when presenting the results of the sensitivity analysis, to focus on the key inference suggested by the model, rather than to confuse the reader with arrays of sensitivity indices relating to intermediate output variables. Piecewise sensitivity analysis, such as when investigating one model compartment at a time, can lead to type II errors if interactions among factors of different compartments are neglected. It is also worth noting that, once a model-based analysis has been produced, most modellers will not willingly submit it to a revision via sensitivity analysis by a third party.

This anticipation of criticism by sensitivity analysis is also one of the 10 commandments of applied econometrics according to Peter Kennedy:

> Thou shall confess in the presence of sensitivity. Corollary: Thou shall anticipate criticism [···] When reporting a sensitivity analysis, researchers should explain fully their specification search so that the readers can judge for themselves how the results may have been affected. This is basically an 'honesty is the best policy' approach, advocated by Leamer, (1978, p. vi) (Kennedy, 2007).

To avoid this pitfall, an analyst should implement uncertainty and sensitivity analyses routinely, both in the process of modelling and in the operational use of the model to produce useful inferences.

Finally the danger of type III error should be kept in mind. Framing error can occur commonly. If a sensitivity analysis is jointly implemented by the owner of the problem (which may coincide with the modeller) and a practitioner (who could again be a modeller or a statistician or a practitioner of sensitivity analysis), it is important to avoid the former asking for just some 'technical help' from the latter upon a predefined framing of the problem. Most often than not the practitioner will challenge the framing before anything else.

1.5 CONCLUDING REMARKS

1. We have just shown different settings for sensitivity analysis, such as:

    - factor prioritization, linked to S_i;
    - factor fixing, linked to S_{Ti} or μ^*;
    - factor mapping, linked to MCF;
    - metamodelling (hints).

CONCLUDING REMARKS

The authors have found these settings useful in a number of applications. This does not mean that other settings cannot be defined and usefully applied.
2. We have discussed fitness for purpose as a key element of model quality. If the purpose is well defined, the output of interest will also be well identified. In the context of a controversy, this is where attention will be focused and where sensitivity analysis should be concentrated.
3. As discussed, a few factors often account for most of the variation. Advantage should be taken of this feature to simplify the results of a sensitivity analysis. Group sensitivities are also useful for presenting results in a concise fashion.
4. Assuming models to be true is always dangerous. An uncertainty/sensitivity analysis is always more convincing when uncertainty has been propagated through more than just one model. Using a parsimonious data-driven and a less parsimonious law-driven model for the same application can be especially effective and compelling.
5. When communicating scientific results transparency is an asset. As the assumptions of a parsimonious model are more easily assessed, sensitivity analysis should be followed by a model simplification.

The reader will find in this and the following chapters didactic examples for the purpose of familiarization with sensitivity measures. Most of the exercises will be based on models whose output (and possibly the associated sensitivity measures) can be computed analytically. In most practical instances the model under analysis or development will be a computational one, without a closed analytic formula.

Typically, models will involve differential equations or optimization algorithms involving numerical solutions. For this reason the best available practices for numerical computations will be presented in the following chapters. For the Elementary Effects Test, we shall offer numerical procedures developed by Campolongo et al. (1999b, 2000, 2007). For the variance-based measures we shall present the Monte Carlo based design developed by Saltelli (2002) as well as the Random Balance Designs based on Fourier Amplitude Sensitivity Test (FAST-RBD, Tarantola et al., 2006, see Chapter 4). All these methods are based on true points in the space of the input factors, i.e. on actual computations of the model at these points. An important and powerful class of methods will be presented in Chapter 5; such techniques are based on metamodelling, e.g. on estimates of the model at untried points. Metamodelling allows for a great reduction in the cost of the analysis and becomes in fact the only option when the model is expensive to run, e.g. when a single simulation of the model takes tens of minutes or hours or more. The drawback is that metamodelling tools such as those developed by Ratto et al. (2007) are less straightforward to encode than plain Monte Carlo. Where possible, pointers will be given to available software.

1.6 EXERCISES

1. Prove that
$$V(Y) = E(Y^2) - E^2(Y).$$

2. Prove that for an additive model of two independent variables X_1 and X_2, fixing one variable can only decrease the variance of the model.
3. Why in μ^* are absolute differences used rather than simple differences?
4. If the variance of Y as results from an uncertainty analysis is too large, and the objective is to reduce it, sensitivity analysis can be used to suggest how many and which factors should be better determined. Is this a new setting? Would you be inclined to fix factors with a larger first-order term or rather those with a larger total effect term?
5. Suppose X_1 and X_2 are uniform variates on the interval [0, 1]. What is the mean? What is the variance? What is the mean of $X_1 + X_2$? What is the variance of $X_1 + X_2$?
6. Compute S_i analytically for model (1.3, 1.4) with the following values: $r = 2, \sigma = \{1, 2\}$ and $\Omega = \{2, 1\}$.
7. Write a model (an analytic function and the factor distribution functions) in which fixing an uncertain factor increases the variance.
8. What would have been the result of using zero-centred distributions for the Ω's in Equation (1.27)?

1.7 ANSWERS

1. Given a function $Y = f(\mathbf{X})$ where $\mathbf{X} = (X_1, X_2, \cdots X_k)$ and $\mathbf{X} \sim p(\mathbf{X})$ where $p(\mathbf{X})$ is the joint distribution of $\mathbf{X}$ with $\int p(\mathbf{X}) d\mathbf{X} = 1$, then the function mean can be defined as

$$E(Y) = \int f(\mathbf{X}) p(\mathbf{X}) d\mathbf{X},$$

and its variance as

$$\begin{aligned}\operatorname{Var}(Y) &= \int (f(\mathbf{X}) - E(Y))^2 p(\mathbf{X}) d\mathbf{X} \\ &= \int f^2(\mathbf{X}) p(\mathbf{X}) d\mathbf{X} + E^2(Y) - 2 \int E(Y) f(\mathbf{X}) p(\mathbf{X}) d\mathbf{X} \\ &= E(Y^2) + E^2(Y) - 2E^2(Y) \\ &= E(Y^2) - E^2(Y).\end{aligned}$$

Using this formula it can easily be proven that $\operatorname{Var}(Y) = \operatorname{Var}(Y + c)$, with c an arbitrary constant. This result is used in Monte Carlo-based variance (and conditional variance) computation by rescaling all values of Y subtracting $E(Y)$. This is done because the numerical error in the variance estimate increases with the value of Y.

2. We can write the additive model of two variables X_1 and X_2 as $Y = f_1(X_1) + f_2(X_2)$, where f_1 is only a function of X_1 and f_2 is only a function of X_2.

Recalling that the variance of (Y) can be written as $V(Y) = E(Y^2) - E^2(Y)$, where E stands for the expectation value, and applying it to Y we obtain

$$V(Y) = E\left(f_1^2 + f_2^2 + 2f_1 f_2\right) - E^2\left(f_1 + f_2\right).$$

Given that $E(f_1 f_2) = E(f_1) E(f_2)$ for independent variables, then the above can be reduced to

$$V(Y) = E\left(f_1^2\right) + E\left(f_2^2\right) - E^2\left(f_1\right) - E^2\left(f_2\right),$$

which can be rewritten as

$$V(Y) = V(f_1) + V(f_2),$$

which proves that fixing either X_1 or X_2 can only reduce the variance of Y.

3. Modulus incremental ratios are used in order to avoid positive and negative values cancelling each other out when calculating the average.

4. It is a new setting. In Saltelli *et al.* (2004) we called it the variance cutting setting, when the objective of sensitivity analysis is the reduction of the output variance to a lower level by fixing the smallest number of input factors. This setting can be considered as relevant in, for example, risk assessment studies. Fixing the factors with the highest total effect term increases our chances of fixing, besides the first-order terms, some interaction terms possibly enclosed in the totals, thus maximizing our chances of reducing the variance (see Saltelli *et al.*, 2004).

5. Both X_1 and X_2 are uniformly distributed in [0, 1], i.e.

$$p(X_1) = p(X_2) = U(0, 1).$$

This means that $p(X_i)$ is 1 for $X_i \in [0, 1]$ and zero otherwise. Thus

$$E(X_1) = E(X_2) = \int_{x=0}^{1} p(x) x \, dx = \left[\frac{x^2}{2}\right]_0^1 = \frac{1}{2}.$$

Further:

$$Var(X_1) = Var(X_2) = \int_{x=0}^{1} p(x) \left(x - \frac{1}{2}\right)^2 dx = \int_{x=0}^{1} \left(x^2 - x + \frac{1}{4}\right) dx$$

$$= [\frac{x^3}{3} - \frac{x^2}{2} + \frac{1}{4}x]_0^1 = \frac{1}{3} - \frac{1}{2} + \frac{1}{4} = \frac{1}{12}$$
$$E(X_1 + X_2) = E(X_1) + E(X_2) = 1,$$

as the variables are separable in the integral.

Given that $X_1 + X_2$ is an additive model (see Exercise 1) it is also true that

$$Var(X_1 + X_2) = Var(X_1) + Var(X_2) = \frac{1}{6}.$$

The same result is obtained integrating explicitly

$$Var(X_1 + X_2) = \int_{x_1=0}^{1} \int_{x_2=0}^{1} p(\mathbf{x})(x_1 + x_2 - 1)^2 \, dx_1 dx_2.$$

6. Note that the model (1.3, 1.4) is linear and additive. Further, its probability density function can be written as the product of the factors' marginal distributions (independent factors). Writing the model for $r = 2$ we have

$$Y(Z_1, Z_2) = \Omega_1 Z_1 + \Omega_2 Z_2$$

with

$$Z_1 \sim N(0, 1) \quad \text{or equivalently} \quad p(Z_1) = \frac{1}{\sigma_{Z_1}\sqrt{2\pi}} e^{-\frac{(Z_1)^2}{2\sigma_{Z_1}^2}}$$

and a similar equation for $p(Z_2)$. Note that by definition the distributions are normalized, i.e. the integral of each $p(Z_i)$ over its own variable Z_1 is 1, so that the mean of Y can be reduced to

$$E(Y) = \Omega_1 \int_{-\infty}^{+\infty} Z_1 p(Z_1) dz_1 + \Omega_2 \int_{-\infty}^{+\infty} Z_2 p(Z_2) dz_2.$$

These integrals are of the type $\int xe^{-x^2} dx$, whose primitive $-e^{-x^2}/2$ vanishes at the extremes of integration, so that $E(Y) = 0$. Given that the model is additive, the variance will be

$$V(Y) = V_{Z_1} + V_{Z_2} = V(\Omega_1 Z_1) + V(\Omega_2 Z_2).$$

For either Z_1 or Z_2 it will be

$$V_{Z_i} = V(\Omega_i Z_i) = \Omega_i^2 V(Z_i).$$

We write

$$V(Z_i) = E(Z_i^2) - E^2(Z_i) = E(Z_i^2)$$

ANSWERS

and

$$E(Z_i^2) = \frac{1}{\sqrt{2\pi}\sigma_{Z_i}} \int_{-\infty}^{\infty} z_i^2 e^{-z_i^2/2\sigma_{Z_i}^2} dz_i.$$

The tabled form is

$$\int_{-0}^{+\infty} t^2 e^{-at^2} dt = \frac{\sqrt{\pi}}{4},$$

which gives with an easy transformation

$$E(Z_i^2) = \sigma_{Z_i}^2$$

so that

$$V_{Z_i} = \Omega_i^2 \sigma_{Z_i}^2$$

and

$$V(Y) = \Omega_1^2 \sigma_{Z_1}^2 + \Omega_2^2 \sigma_{Z_2}^2$$

and

$$S_{Z_i} = \frac{\Omega_i^2 \sigma_{Z_i}^2}{V(Y)}.$$

Inserting the values $\sigma = \{1, 2\}$ and $\Omega = \{2, 1\}$ we obtain $V(Y) = 8$ and $S_{Z_1} = S_{Z_2} = \frac{1}{2}$.

The result above can be obtained by explicitly applying the formula for S_i to our model:

$$S_{Z_i} = \frac{V(E(Y \mid Z_i))}{V(Y)},$$

which entails computing first $E(Y \mid Z_i = z_i^*)$. Applying this to our model $Y = \Omega_1 Z_1 + \Omega_2 Z_2$ we obtain, for example, for factor Z_1:

$$E(Y \mid Z_1 = z_1^*) = \int_{-\infty}^{+\infty} p(z_1)p(z_2)\,(\Omega_1 z_1^* + \Omega_2 z_2)\,dz_1 dz_2 = \Omega_1 z_1^*.$$

Hence V_{Z_1} – the variance over z_1^* of $\Omega_1 z_1^*$ – is, as before, equal to $\Omega_1^2 \sigma_{Z_1}^2$ and

$$S_{Z_i} = \frac{\Omega_i^2 \sigma_{Z_i}^2}{\Omega_1^2 \sigma_{Z_1}^2 + \Omega_2^2 \sigma_{Z_2}^2}.$$

7. We consider the model $Y = X_1 \cdot X_2$, with the factors identically distributed as

$$X_1, X_2 \sim N(0, \sigma).$$

Based on the previous exercise it is easy to see that

$$E(Y) = E(X_1 X_2) = E(X_1)E(X_2) = 0,$$

so that

$$V(Y) = E\left(X_1^2 X_2^2\right) = E\left(X_1^2\right) E\left(X_2^2\right) = \sigma^4.$$

If X_2 is fixed to a generic value x_2^*, then

$$E(X_1 x_2^*) = x_2^* E(X_1) = 0$$

as in a previous exercise, and

$$V(Y \mid X_2 = x_2^*) = V(X_1 x_2^*) = E\left(X_1^2 (x_2^*)^2\right) = (x_2^*)^2 E\left(X_1^2\right) = (x_2^*)^2 \sigma^2.$$

It is easy to see that $V(X_1 x_2^*)$ becomes bigger than $V(Y)$ whenever the modulus of x_2^* is bigger than σ.

Further, from the relation

$$V(Y \mid X_2 = x_2^*) = (x_2^*)^2 \sigma^2$$

one gets

$$E(V(Y \mid X_2)) = \sigma^4 = V(Y).$$

Given that

$$E(V(Y \mid X_2)) + V(E(Y \mid X_2)) = V(Y)$$

it must be that

$$V(E(Y \mid X_2)) = 0,$$

i.e. the first-order sensitivity index is null for both X_1 and X_2. These results are illustrated in the two figures which follow.

Figure 1.13 shows a plot of $V_{X_{\sim 2}}(Y \mid X_2 = x_2^*)$, i.e. $V_{X_1}(Y \mid X_2 = x_2^*)$ at different values of x_2^* for $\sigma = 1$. The horizontal line is the unconditional variance of Y. The ordinate is zero for $x_2^* = 0$, and becomes higher than $V(Y)$ for $x_2^* \sim 1$.

Figure 1.14 shows a scatterplot of Y versus x_1^* (the same shape would appear for x_2^*). It is clear from the plot that whatever the value of

ANSWERS

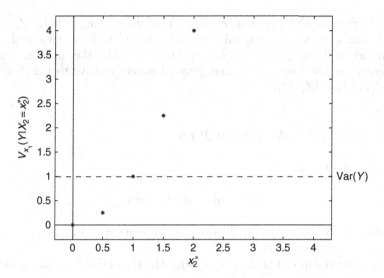

Figure 1.13 $V_{X_1}(Y \mid X_2 = x_2^*)$ at different values of x_2^*

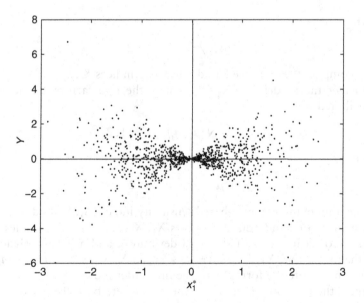

Figure 1.14 Scatterplot of Y versus x_1^*

the abscissa, the average of the points on the ordinate is zero, i.e. $E_{\mathbf{X}_{-1}}(Y \mid X_1) = E_{X_2}(Y \mid X_1) = 0$. It is also clear from Figure 1.14 that even $V_{X_1}\left(E_{X_2}(Y \mid X_1)\right)$ will be zero, such that both S_1 and S_2 are zero for this model and all variance is captured by the second-order term, i.e. $S_{12} = 1$.

8. Referring to the previous exercise it is clear that if both Z_i and Ω_i are centred in zero, all first-order terms will be zero and the model will be purely interactive. In this case the only nonzero terms are the four interactions (second order) relative to the couples $(Z_1, \Omega_1), \cdots (Z_4, \Omega_4)$.

1.8 ADDITIONAL EXERCISES

1. Given the function

$$f(x) = \sin(X_1 \sin(X_2 \sin(X_3)))$$

   with X_1, X_2, X_3 distributed normally with mean zero, can you guess what the first-order indices will be?

2. Consider the model $\Omega_1 Z_1 + \Omega_2 Z_2$ with Ω_1, Ω_2 as fixed constants and

$$Z_i \sim N(\mu_{Z_i}, \sigma_{Z_i}), \quad i = 1, 2$$

   with

$$\mu_{Z_i} \neq 0, \quad i = 1, 2$$

   and compute the variance-based sensitivity indices S_1, S_2.

3. Consider the model $Y = X_1 \cdot X_2$, where the two factors are normally distributed as

$$X_i \sim N(\mu_i, \sigma_i), \quad i = 1, 2$$

   with

$$\mu_i \neq 0, \quad i = 1, 2$$

   and compute the variance-based sensitivity indices S_1, S_2, and S_{12}.

4. Given a set of standardized variables $X_1, X_2, \ldots, X_k$ (all variables have thus zero mean and unit standard deviation), and a linear polynomial of the form $f(X_1, X_2, \ldots, X_k) = a_0 + \sum_{i=1}^{k} a_i X_i$, where $a_0, a_1, \cdots, a_k$ are constants, write the formula for the first-order indices S_i.

5. Repeat the previous exercise, for the case where both the $a_0, a_1, \cdots, a_k$ and the $X_1, X_2, \ldots, X_k$ are normally distributed:

$$X_i \sim N(\mu_{X_i}, \sigma_{X_i})$$

   and

$$a_i \sim N(\mu_{a_i}, \sigma_{a_i}).$$

1.9 SOLUTIONS TO ADDITIONAL EXERCISES

1. The first-order indices will be zero.
2. The solution is

$$S_{Z_i} = \frac{\Omega_i^2 \sigma_{Z_i}^2}{V(Y)}$$

as for the case with

$$\mu_{Z_i} = 0, \quad i = 1, 2$$

3. The solution is

$$S_1 = \frac{\mu_2^2 \sigma_1^2}{\left(\mu_1^2 \sigma_2^2 + \mu_2^2 \sigma_1^2 + \sigma_1^2 \sigma_2^2\right)}$$

and analogous formula for S_2, while

$$S_{12} = \frac{\sigma_1^2 \sigma_2^2}{\left(\mu_1^2 \sigma_2^2 + \mu_2^2 \sigma_1^2 + \sigma_1^2 \sigma_2^2\right)}.$$

4. It is simply

$$S_i = \frac{a_i^2}{\sum_{i=1}^{k} a_i^2},$$

i.e. each sensitivity index is proportional to the square of its coefficient.

5. The problem is additive in a_0 and in the k sets $\{a_i, X_i\}$. Using this and the results from Exercise 3 it is easy to derive the solution.

$$S_{a_i} = \frac{\mu_{X_i}^2 \sigma_{a_i}^2}{V}$$

$$S_{X_i} = \frac{\mu_{a_i}^2 \sigma_{X_i}^2}{V}$$

$$S_{a_i X_i} = \frac{\sigma_{a_i}^2 \sigma_{X_i}^2}{V}$$

$$V = \sum_{i=1}^{k} \left(\mu_{a_i}^2 \sigma_{X_i}^2 + \mu_{X_i}^2 \sigma_{a_i}^2 + \sigma_{a_i}^2 \sigma_{X_i}^2\right).$$

By putting $a_0 = 0$ the above solution can be used to compute the sensitivity indices for model (1.3, 1.27).

2

Experimental Designs

2.1 INTRODUCTION

WHEN HAVE RESEARCHERS DONE SIMILAR EXPERIMENTAL STUDIES IN THE PAST?
HOW DOES SENSITIVITY ANALYSIS DIFFER FROM PREVIOUS STUDIES?

In 1948, the National Heart Institute enlisted the help of over 5000 residents of the town of Framingham, Massachusetts. Researchers interviewed the volunteers and examined them physically, recording numerous details about their health and lifestyles; they subsequently revisited Framingham every two years for follow-up examinations. The purpose of the study was to investigate the incidence of heart disease and to determine the circumstances that gave rise to it. The pioneering Framingham Heart Study (NIH, 2002) revealed connections between lifestyle choices (smoking, physical activity), medical conditions (e.g. high blood pressure) and the contraction of heart disease.

The scale of the Framingham study doubled in 1971 when a new generation of descendants was incorporated into the investigation, and other health studies have used even larger groups of people. For instance, the Harvard Nurses Health Study (Hankinson, 2001) tracks over 100 000 women in an investigation of the effects of diet on the development of chronic diseases. Such large studies can screen many factors and many combinations of factors. They can also discriminate among factors (e.g. which is more effective in avoiding heart attacks: losing weight or taking up physical activity?).

Global Sensitivity Analysis. The Primer A. Saltelli, M. Ratto, T. Andres, F. Campolongo, J. Cariboni, D. Gatelli, M. Saisana and S. Tarantola © 2008 John Wiley & Sons, Ltd

However, even very large general studies are not appropriate to obtain certain kinds of information, such as the best treatment of very rare diseases or the effectiveness of experimental as opposed to standard treatments. In considering rare diseases, a large study may not contain enough people suffering from the disease to get good statistical results. In determining the effectiveness of new treatments there are more powerful techniques available, such as pairing individuals with similar conditions and lifestyles. In a double-blind experiment, researchers give one of the pair the standard treatment, and the other the experimental treatment, without anyone knowing who has received which. A study of this kind may require only a few dozen individuals to determine whether the new treatment is promising enough to merit further study.

In carrying out sensitivity analysis of simulation models researchers are faced with similar choices. A large randomized study can answer most sensitivity analysis questions, but at a large cost in terms of computer execution time and data management. In contrast, careful use of experimental designs can often answer more specific questions, with an investment of fewer resources.

In recent years, companies with large databases of customers and sales have often benefited from investigating possible connections among the fields of the database. A new discipline called *data mining* has emerged to organize research in this area. Data mining techniques can be applied to a simulation data set generated in a randomized fashion.

By contrast, sensitivity analysis tends to focus more on the application of a suitable experimental design to extract specific types of data. The strength of sensitivity analysis techniques is that they can be tuned to elicit information not readily available to data mining techniques.

Sensitivity analysis uses and extends experimental techniques originally developed for scientific research. By the early twentieth century, experimental research had become quantitative and comparative, teasing out experimental results from raw data using powerful statistical techniques. It was at this time that statisticians like Fisher and 'Student' (William Sealey Gosset) developed the theory of distributions like Student's-t and the chi-square distribution, useful in statistical testing.

In the design of experiments, a researcher has several variables that can be controlled (e.g. temperature, velocity, concentration, duration) There may be other variables that cannot be controlled but which can affect the results of an experiment (such as air pressure and humidity, genetic effects, unknown material defects). The goal is to determine how and how much each variable affects one or more measurements.

The set-up of simulation results for sensitivity analysis can be quite similar to the set-up for physical experimentation, so the results of the last century of experimental design are applicable. But there are also some differences that lead us into experimental regimes that would not be feasible for physical experimentation.

Experimental simulations allow us, for example, to explore the behaviour of complex systems. There may be many more variables involved than a physical experimenter would be able to handle, especially when exploring law-driven models. (For example, researchers could use simulations to investigate possible environmental effects of toxic materials.) Complex systems can have physical, chemical, biological and environmental components, all acting in concert and influencing one another.

To understand the effect of parameters in complex systems, we must first examine their effects in simple systems.

2.2 DEPENDENCY ON A SINGLE PARAMETER

How do we characterize a single parameter? What standards should we follow in designating its probability distribution? How do we sample values for parameters with different distributions? What possible effects could a single parameter have?

If an output variable depends on only a single parameter, one might think that a sensitivity analysis would be straightforward. It is true that identifying the influential parameter is a minor matter. In the notation of Chapter 1, there is only one sensitivity coefficient $S_1 = 1$, indicating that X_1 accounts for all the variance in the output. Nevertheless, characterizing the influence can still be a challenge. Some simple strategies for dealing with the single-parameter case can be generalized to sensitivity analysis of larger numbers of parameters. Conversely, sensitivity analysis of a complex model may degenerate to a study of the single dominant parameter. A multivariate approach should be able to generate useful information even in this constricted case.

Without loss of generality, call the single parameter X_1. In later sections we will expand the discussion to cover a set of k parameters $\{X_r\}_{r=1,k}$ that can affect an output simultaneously.

Assume that X_1 has a finite domain, from 0 to 1. This is one way of characterizing the domain of X_1, of several that have been used in the past. For instance, in cases where two-level designs are employed, it is traditional to designate the domain of a parameter to be $[-1, 1]$, and this convention will be followed in Section 2.4. The two discrete levels used in this type of design are designated ± 1. If the design also uses a midpoint for X_1, 0 represents the midpoint. In other designs with $s > 2$ discrete levels, it is customary to designate the levels as $0, 1, \ldots s-1$, and this approach will also be used in discussing Latin hypercube sampling. Eventually, however, samples for X_1 will be transformed to the standard continuous domain $[0, 1]$.

If we begin the analysis with a variable that has a finite domain other than [0, 1], or even an infinite domain, we can normalize it by a transformation, so that the normalized value lies in this interval. The transformation can also remove a physical unit, like metres or degrees, converting the parameter to a standard form.

For instance, let parameter Z_1 be a continuous random variable, with a probability density function (pdf) $f_{Z_1}(z)$, and a cumulative distribution function (cdf) $F_{Z_1}(z) = \int_{-\infty}^{z} f_{Z_1}(z')dz'$. Let X_1 be a transformed variable based on Z_1 according to the following equation:

$$X_1 = F_{Z_1}(Z_1). \qquad (2.1)$$

Then the domain of X_1 is the interval [0, 1], by the definition of cdf. While Z_1 has an arbitrary domain and physical unit, X_1 is restricted to [0, 1], and as a cumulative probability, is unitless. Moreover, we can assume that X_1 is uniformly distributed across the interval, rather than concentrated in one part of the interval. This again follows from the transformation in Equation (2.1). Whatever distribution Z_1 has, when X_1 is defined as the value of Z_1's cdf, X_1 automatically acquires a uniform distribution from 0 to 1, since

$$F_{X_1}(x) = \Pr\left\{F_{Z_1}(Z_1) \leqslant x\right\} = x \qquad (2.2)$$

in this interval.

Given a normalized parameter value X_1 between 0 and 1, an analyst can determine the associated value for Z_1 through the inverse transformation

$$Z_1 = F_{Z_1}^{-1}(X_1). \qquad (2.3)$$

If X_1 is uniformly distributed over [0, 1], then Z_1 will follow the distribution specified by $F_{Z_1}(z)$. For many standard distributions, the mapping between Z_1 and X_1 can be inverted in this way without ambiguity. Therefore, since we can use normalized variables in an analysis if necessary, we will simply assume that our single parameter X_1 is uniformly distributed in the unitless interval [0, 1].

To analyse the effect of this single, uniformly distributed parameter X_1, an analyst chooses an experimental design that will extract the maximum information about the influence of X_1 on a simulation outcome Y. Standard strategies assume the model is a black box, and that the only way of extracting information is through simulations. If other sources of information are available (e.g. the slope of the response curve, found by taking derivatives), then they may be used to modify the strategy.

What possible types of influence could X_1 have on Y? Consider Figure 2.1. It shows a few of the infinite possibilities. In each panel of the figure, Y is plotted as a function of X_1. Figure 2.1(a) shows the influence of a

DEPENDENCY ON A SINGLE PARAMETER

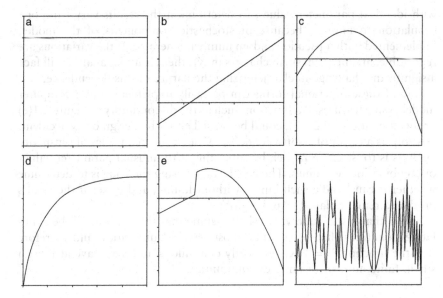

Figure 2.1 Examples of single-variable functions

parameter that has no effect at all on Y. In large models, there may be many parameters like this. They appear in a model for completeness, and they may affect some other output, but not the output of interest. Alternatively, the parameter's distribution may be so narrow that its possible variation, if any, has little effect on the output of interest. The variable X_1 could even be the transformed representation of a constant parameter Z_1, so that all values of X_1 between 0 and 1 correspond to a single fixed value of Z_1.

Figures 2.1(b)–(d) show common forms of smooth dependencies. Figures 2.1(b) and (c) are respectively linear and quadratic polynomials. The former is completely determined by two points, and its extreme values occur at the endpoints of the interval. The quadratic polynomial is completely determined by three points. Extreme values of a nonlinear curve may occur as a local extremum within the interval, like the maximum in Figure 2.1(c), or at an endpoint, like the minimum value in the same figure. Note that in Figure 2.1(c), the initial slope of the curve is opposite in sign to the average slope one would deduce from the two endpoints alone. In contrast to these two polynomials, Figure 2.1(d) shows an asymptotic curve that cannot easily be represented by a low-order polynomial. It has endpoint extrema, and most of the variation occurs over the left half of its domain.

Figure 2.1(e) and (f) show discontinuous functional forms. Figure 2.1(f) is highly variable, and could represent a signal affected by a large amount of noise. In that case it would be worth repeating some of the simulations with identical parameter values, to determine if the results vary from one simulation to another because of stochastic components of the model, implemented with a pseudo-random number generator. If the variations are repeatable, arising from small changes in X_1, the sensitivity analyst will face a significant challenge in characterizing the nature of this dependence.

Any of these functional forms can be easily implemented in a computer model. Some features of a function, such as the discontinuity in Figure 2.1(e) or the oscillations in 2.1(f), could be caused either by design or by accident, possibly as a result of a software defect. One of the main roles of sensitivity analysis is to 'stress' a model, by subjecting it to unusual parameter values or combinations of values. The benefit of stressing the code is to determine whether it can handle such inputs without failing, and to see if the results produced remain plausible and realistic.

In the analysis to follow, we shall assume that the analyst will be simulating quantities that behave like most physical quantities, and therefore that the function $Y = Y(X_1)$ is mostly continuous and may have at most a small, finite number of jump discontinuities.

2.3 SENSITIVITY ANALYSIS OF A SINGLE PARAMETER

HOW DO WE PERFORM SENSITIVITY ANALYSIS ON A SINGLE PARAMETER? HOW SHOULD PARAMETER VALUES BE SELECTED? WHAT IS STRATIFIED SAMPLING? HOW DO DIFFERENT SAMPLING METHODS AFFECT THE ESTIMATION OF OUTPUT MEAN AND VARIANCE?

How much information a sensitivity analysis will reveal about a parameter's influence depends on the number of sample points that are simulated and where they are located.

2.3.1 Random Values

If the data come from a random source (e.g. random samples from a database or a pseudo-random number generator), they will be scattered in an uncontrolled manner across the domain of X_1. Suppose the data points are $\{(x_{i1}, y_i)\}_{i=1,N}$, where N is the number of points in the sample. There is a problem with such data, which can readily be corrected by experimental design. In particular, the values for X_1 are not evenly distributed across the interval [0, 1]. The density of points in one part of the interval can differ from that in another. In fact, if examined closely, the values for X_1 exhibit *clusters* and *gaps*. Clusters occur when several values are quite close together. Gaps are regions without any points at all.

SENSITIVITY ANALYSIS OF A SINGLE PARAMETER

As N grows large the density evens out and larger gaps will be filled in, while smaller gaps remain. For this reason, random samples for X_1 give rise to unbiased estimates of the mean and variance of Y:

$$\bar{Y} = \frac{1}{N}\sum_{i=1}^{N} y_i$$

$$\hat{V}_Y = \frac{1}{N-1}\sum_{i=1}^{N}(y_i - \bar{Y})^2. \qquad (2.4)$$

However, a mean and variance estimated this way are uncertain. For instance, the uncertainty in the mean estimate $\bar{Y}$ can be characterized by its standard error $s_{\bar{Y}}$:

$$s_{\bar{Y}} = \sqrt{\frac{1}{N^2}\sum_{i=1}^{N}\mathrm{Var}\{y_i\}} \cong \sqrt{\frac{\hat{V}_Y}{N}}. \qquad (2.5)$$

Because the standard error depends on the square root of N, the uncertainty shrinks slowly as N increases.

2.3.2 Stratified Sampling

Most designs use some sort of stratified sampling to improve the rate at which estimated quantities converge to the true quantities. In stratified sampling, the domain of X_1 is divided into subintervals (often of equal length), and the sampling is constrained so that each subinterval contains the same number of sample points. The points themselves may be selected systematically to lie at particular locations within these subintervals, or they may be sampled randomly within each subinterval.

Figure 2.2 shows several number lines illustrating different ways of distributing 16 points across the interval from 0 to 1. The dashed vertical

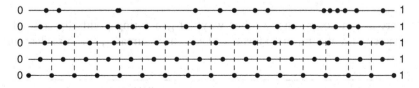

Figure 2.2 Different ways of distributing points from 0 to 1

lines show subinterval boundaries when the interval is subdivided. The distributions of points in the figure have the following properties:

- Top line: independently and identically distributed random points, with visible gaps and clusters. Some of the points overlap in the figure. This type of sample is easy to generate with a pseudo-random generator. By Equation (2.4), it yields unbiased estimates of the mean and variance of Y.
- Second line: a stratified sample, with eight subintervals, and two points randomly selected within each subinterval. Equation (2.4) can be applied to each subinterval to yield unbiased mean and variance estimates. These can be combined under the assumption that each subinterval has equal probability, yielding unbiased mean and variance estimates for Y covering the entire interval. It is also possible to compare variance *among* subintervals against variance *within* subintervals to determine if parameter X_1 has a significant influence.
- Middle line: a stratified sample, with 16 subintervals, and one point randomly selected within each subinterval. Equation (2.4) can be applied to each subinterval to yield unbiased mean estimates. The variance among intervals can be estimated, but not the variance within intervals, as there is only one point per subinterval. (However, each pair of adjacent intervals can be combined to reduce the situation to the previous case.)
- Fourth line: a stratified sample, with 16 intervals, and one point at each interval's midpoint. Using a trapezoidal approximation (i.e. fitting a bar graph to the data), a good estimate of the area under the curve $Y = Y(X_1)$ in each subinterval can be computed. From this area, a mean estimate can be produced, but it will be biased[1] by the selection of subinterval midpoints. A variance estimate among subintervals can be made, but it is also biased.
- Bottom line: a stratified sample, with 16 subintervals, and a point at the ends of each subinterval, for a total of 17 points instead of 16. By joining adjacent (x_{i1}, y) points with a straight line, an analyst can compute a good estimate of the area under the curve $Y = Y(X_1)$ in each subinterval, yielding a (biased) mean estimate. By using a straight line approximation to the curve in each interval, (biased) variance estimates among and within subintervals can be obtained.

There are of course many other ways of generating points for a single parameter, and of computing mean and variance estimates, but this selection

---

[1] A *biased* estimator has an expected value that is potentially different from the true value. Redoing the analysis many times does not help to find the true value.

SENSITIVITY ANALYSIS OF A SINGLE PARAMETER

provides a good starting point. The following paragraphs describe how the computations can be done.

2.3.3 Mean and Variance Estimates for Stratified Sampling

Suppose the N points $\{(x_{i1}, y_i)\}_{i=1,N}$ come from $m \leq N$ subintervals of equal length, where m divides into N. Assume X_1 is uniformly distributed, so that these subintervals have equal probability. Assume further that there may be another source of variation beside X_1 affecting the output Y. For example, the model may include stochastic behaviour, or other parameters may change value from run to run in an unspecified manner.

By applying Equation (2.4) to each subinterval separately, we can devise m mean and variance estimates $\bar{Y}_j$ and $\hat{V}_{Y_j}$

$$\bar{Y}_j = \frac{1}{N/m} \sum_i y_i \qquad \hat{V}_{Y_j} = \frac{1}{(N/m) - 1} \sum_i (y_i - \bar{Y}_j)^2 \qquad (2.6)$$

where in each summation, i varies over points belonging to subinterval j.

The variance estimate in Equation (2.6) applies only when $m < N$. When these quantities are equal, N/m equals one point per subinterval, and the variance calculation is invalid as it would involve division by zero.

The global mean and variance when N/m parameter values are randomly sampled from each subinterval are calculated as shown in the following equations. Note that the mean estimator is still the overall sample mean.

$$\bar{\bar{Y}} = \frac{1}{m} \sum_{j=1}^{m} \bar{Y}_j = \frac{1}{N} \sum_{i=1}^{N} y_i \qquad (2.7)$$

$$\tilde{V}_Y^a = \frac{N/m}{m-1} \sum_{j=1}^{m} (\bar{Y}_j - \bar{\bar{Y}})^2 \qquad (2.8)$$

$$\tilde{V}_Y^w = \frac{1}{m} \sum_{j=1}^{m} \hat{V}_{Y_j}. \qquad (2.9)$$

The two variance estimates have superscripts a and w, indicating that they are based on the variances *among* and *within* subintervals respectively. If Y does not depend on X_1, these quantities are independent estimates of the variance of Y. If Y does depend on X_1, the ratio of one to the other is a measure of the degree of influence of X_1.

These equations apply when the points in an interval are randomly selected. Suppose instead that there is one deterministic value in the centre of the jth subinterval, at $x_{j1} = (2j-1)/2N$, for j from 1 to N, as in the fourth line of Figure 2.2. Then estimates of the mean and variance must be based on some model of variation within the interval. The simplest assumption is that in the jth interval the function $Y(X_1)$ is linear, with an unknown slope a_j:

$$y = y_j + a_j(x - x_{j1}). \qquad (2.10)$$

Because the sample point is in the middle of the subinterval, contributions from the left and right sides of the subinterval cancel each other out, and the average value of y in the jth interval for a linear fit is

$$\bar{Y}_j = \frac{1}{1/N} \int_{x_{j-1,1}}^{x_{j1}} (y_j + a_j(x - x_{j1})) \, dx \qquad (2.11)$$
$$= y_j$$

irrespective of the value of a_j.

The diagram on the right shows several different trapezoids with different slopes, all passing through the same point in the middle of an interval. The area of each trapezoid (i.e. the width by the average height) is the same. The average value in Equation (2.11), which corresponds to the average height of each trapezoid, is the same. However, each of these trapezoids would have a different variance of heights in the interval, so it is not feasible to estimate $\hat{V}_{Y_j}^w$. Equations (2.7) and (2.8) still apply.

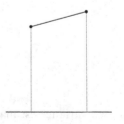

If we use endpoint values for each interval (as in the bottom line of Figure 2.2), the simplest model of the function $Y = Y(X_1)$ is the line that interpolates the two endpoints, as shown on the right. One extra point (x_{01}, y_0) is required for X_1 at the left end of the first subinterval.

This is not a bad approximation, as shown in Figure 2.3, where the nonpolynomial functions from Figure 2.1 have been interpolated at 11 evenly spaced points, and joined by straight lines. A sequence of connected straight lines like this is called a *piecewise linear* fit. In the last frame of Figure 2.3, it is clear that many more interpolation points would be required to fit this jagged function, but in the other two panels curves and discontinuities have been handled fairly well.

SENSITIVITY ANALYSIS OF A SINGLE PARAMETER

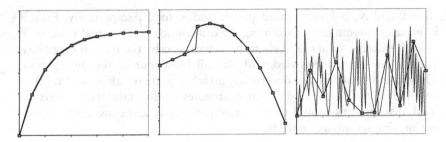

Figure 2.3 Piecewise linear fits to nonpolynomial functions

In this case the linear interpolating function in the jth subinterval for parameter X_1 is

$$y = y_j + \frac{(x - x_{j1})}{(x_{j-1,1} - x_{j1})}(y_{j-1} - y_j) \tag{2.12}$$

for j from 1 to k. The average value is

$$\overline{Y}_j = \frac{1}{1/N} \int_{x_{j-1,1}}^{x_{j1}} \left(y_j + \frac{(x - x_{j1})}{(x_{j-1,1} - x_{j1})}(y_{j-1} - y_j) \right) dx$$

$$= \frac{1}{2}(y_{j-1} + y_j) \tag{2.13}$$

and the estimated variance of Y in the jth interval is

$$\hat{V}_{Y_j} = \frac{1}{1/N} \int_{x_{j-1,1}}^{x_{j1}} \left(y_j + \frac{(x - x_{j1})}{(x_{j-1,1} - x_{j1})}(y_{j-1} - y_j) - \overline{Y}_j \right)^2 dx$$

$$= \frac{1}{12}(y_j - y_{j-1})^2 \tag{2.14}$$

which is the variance of a uniform variate between y_{j-1} and y_j.

With these definitions the variance equations (2.8) and (2.9) apply. The overall estimate of the mean is slightly different in that the endpoints are given half as much weight as the other points.

$$\overline{Y} = \frac{1}{N} \sum_{j=1}^{N} \overline{Y}_j = \frac{1}{2N}(y_0 + 2y_1 + 2y_2 + \cdots + 2y_{N-1} + y_N). \tag{2.15}$$

One disadvantage of using endpoints is that the extreme values $X_1 = 0$ and $X_1 = 1$ can cause problems. Suppose X_1 is a transformed version of a normal (Gaussian) variate Z_1, as defined by Equation (2.1). Then points

$X_1 = 0$ and $X_1 = 1$ correspond to $-\infty$ and ∞ for Z_1 respectively. Even if infinities are avoided by truncating the distribution,[2] the actual value of Y at such extreme values is likely to be determined by the truncation points, and by our limited knowledge of the tail behaviour of the distribution. There could be significant uncertainty attached to these values, and yet they could affect both mean and variance estimates significantly. If endpoints are used, distributions should be truncated judiciously and consistently before performing sensitivity analysis.

2.4 SENSITIVITY ANALYSIS OF MULTIPLE PARAMETERS

HOW DO WE SAMPLE MANY PARAMETERS SO AS TO OBTAIN INFORMATION ABOUT THE EFFECT OF EACH, OR AT LEAST OF THE MOST INFLUENTIAL? WHAT TYPES OF ANALYSIS SHOULD A SAMPLING SCHEME SUPPORT? HOW MANY INFLUENTIAL PARAMETERS DO WE EXPECT TO FIND? WHAT ARE ONE-AT-A-TIME DESIGNS AND FRACTIONAL FACTORIAL DESIGNS? WHAT IS QUASI-RANDOM SAMPLING?

Suppose the analyst has the capability to manipulate a set of k parameters, $\{X_r\}_{r=1,k}$. This is the situation discussed in general terms in Chapter 1. Each parameter has properties like those described in the last section for X_1. Specifically, the domain of X_r is the unitless interval from 0 to 1. Each X_r is uniformly distributed across this domain, possibly because it is the transformed version of some variate Z_r.

For now, assume that the parameters are all statistically independent. That is, knowing specific values for some subset of the parameters says nothing about the distribution of the remaining parameters. That would *not* typically be the case if there were a constraint linking the parameters. As an example of a constraint, suppose that we know $\sum_{r=1}^{k} X_r = k/2$ exactly. The expected value of the sum of k independent parameters is $k/2$, so this constraint is consistent with what we know about each parameter. Should the situation arise where we have specified low values for half the parameters, however, this constraint would limit our freedom to select the rest. There are ways of dealing with such a situation, but sensitivity analysis is much simpler if we avoid it in the first place. In this case, we

---

[2] Truncating the distribution means limiting the lower and upper values of Z_1 to some finite values, such as ± 3 standard deviations from the mean. There would be only about a quarter of 1% chance of sampling values further out than that for a normal variate. The modified cdf for Z_1 after truncation would be $F'_{Z_1}(z) = \int_{\mu-3\sigma}^{z} f_{Z_1}(z')dz' \Big/ \int_{\mu-3\sigma}^{\mu+3\sigma} f_{Z_1}(z')dz'$, which means that sampled points from the truncated distribution using Equation (2.3) would all be slightly different from those sampled from the original distribution.

SENSITIVITY ANALYSIS OF MULTIPLE PARAMETERS

could change one of the parameters from a parameter to a derived quantity $X_n = k/2 - \sum_{r \neq n} X_r$ to relieve pressure on the sampling.[3]

In the following discussion, we will assume that any combination of parameter values $\{x_r\}_{r=1,k}$ is equally likely. The k sampled parameter values define a point in k-dimensional space, and sample points are uniformly distributed throughout a k-dimensional hypercube of side 1.

2.4.1 Linear Models

As in the analysis of a single parameter, the choice of experimental design depends on how the experimenter expects the parameters to affect the model output. As a first approximation, the dependency can be viewed as linear in each parameter. That is, if Y is the output, and the inputs are the k variables X_1 to X_k, the entire model being simulated would behave approximately like so:

$$Y = b_0 + \sum_{r=1}^{k} b_r X_r$$
$$= b_0 + b_1 X_1 + b_2 X_2 + \cdots + b_{k-1} X_{k-1} + b_k X_k \quad (2.16)$$

where the b_r's are all constants that we assume are unknown at the start of sensitivity analysis.

When the model is run with a set of parameter values, a data point becomes available for sensitivity analysis. Any set of N simulations, where $N \geq k+1$, can be solved for the b_r's, provided that the system of equations that is produced is linearly independent (e.g. there can be no repeated simulations). In general, N simulations will result in the following $N \times (k+1)$ system of linear equations,

$$\begin{bmatrix} 1 & x_{11} & \cdots & x_{1k} \\ 1 & x_{21} & \cdots & x_{2k} \\ \vdots & \vdots & \ddots & \vdots \\ 1 & x_{N1} & \cdots & x_{Nk} \end{bmatrix} \begin{pmatrix} b_0 \\ b_1 \\ \vdots \\ b_k \end{pmatrix} = \begin{pmatrix} y_1 \\ y_2 \\ \vdots \\ y_N \end{pmatrix} \quad (2.17)$$

which we can abbreviate as

$$X_{Nk} B_k = Y_N \quad (2.18)$$

using matrix notation.

---

[3] Note that X_n would not necessarily lie in the interval [0, 1].

The matrix X_{Nk} has 1's in the first column, and experimental values for the k parameters in the N simulations in the remaining columns. B_k contains the $k+1$ unknown coefficients corresponding to the intercept b_0 and the k parameters. Y_N contains the N output values from the N simulations.

If $N \geqslant k+1$, it is generally possible to solve these equations. If N is strictly greater than $k+1$, it will not be possible to solve the equations exactly, unless the model is in fact linear, as the system of equations is overdetermined. However, a solution of the least-squares type will generally be available. Solving an overdetermined system of equations for a least-squares solution is computationally intensive. Numerical problems can also arise if the equations are ill-conditioned (e.g. if several points are quite close together). Random samples (i.e. a sample where every value in X_{Nk} outside the first column is sampled randomly) tend to be poorly conditioned for large k because of clustering. Nevertheless, off-the-shelf software can often solve such systems. Even general purpose spreadsheet packages like Microsoft's Excel have regression equation solvers that can be used to determine good estimates of the b_r coefficients, provided there are not too many of them.

With only a small number of parameters, a reasonably good sensitivity analysis can sometimes be carried out through a regression analysis alone. After performing an initial regression using Equation (2.17) and random or systematically sampled parameter values, the analyst can look at the residuals (i.e. the discrepancies between the actual values for Y, and the values estimated using the linear model). The residuals may exhibit a pattern that suggests additional terms for the regression model, such as quadratic terms in one or more of the parameters.[4]

If $N < k+1$, or if the system has dependencies that reduce its effective size (e.g. if a simulation is repeated two or more times), there will be an infinite number of solutions to the equations. However, that does not mean that the equations are devoid of useful information, as we shall see later when discussing group sampling.

2.4.2 One-at-a-time (OAT) Sampling

A sensitivity analyst can avoid or limit numerical problems in solving Equation (2.17) by carefully selecting the data points that define X_{Nk}. One way of simplifying X_{Nk} is to use a 'one-at-a-time' (OAT) design, where only

---

[4] In this exposition there is no need to resort to the standardized regression coefficients mentioned in Chapter 1, since the assumption that every parameter has the same uniform distribution eliminates differences among them that can be clarified by use of SRCs.

SENSITIVITY ANALYSIS OF MULTIPLE PARAMETERS

one parameter changes values between consecutive simulations. Suppose, for example, that Equation (2.17) is rewritten as

$$\begin{bmatrix} 1 & 0 & 0 & 0 & 0 & \cdots & 0 \\ 1 & 1 & 0 & 0 & 0 & \cdots & 0 \\ 1 & 1 & 1 & 0 & 0 & \cdots & 0 \\ 1 & 1 & 1 & 1 & 0 & \cdots & 0 \\ 1 & 1 & 1 & 1 & 1 & \cdots & 0 \\ \vdots & \vdots & \vdots & \vdots & \vdots & \ddots & \vdots \\ 1 & 1 & 1 & 1 & 1 & \cdots & 1 \end{bmatrix} \begin{pmatrix} b_0 \\ b_1 \\ \vdots \\ b_k \end{pmatrix} = \begin{pmatrix} y_1 \\ y_2 \\ \vdots \\ y_{k+1} \end{pmatrix} \qquad (2.19)$$

where every variable takes only two values, 0 and 1, and only one variable changes its value between each pair of consecutive simulations. This system of equations can be simplified by elementary row operations. In every row but the first, subtract the entries from the previous row, yielding

$$\begin{bmatrix} 1 & 0 & 0 & 0 & 0 & \cdots & 0 \\ 0 & 1 & 0 & 0 & 0 & \cdots & 0 \\ 0 & 0 & 1 & 0 & 0 & \cdots & 0 \\ 0 & 0 & 0 & 1 & 0 & \cdots & 0 \\ 0 & 0 & 0 & 0 & 1 & \cdots & 0 \\ \vdots & \vdots & \vdots & \vdots & \vdots & \ddots & \vdots \\ 0 & 0 & 0 & 0 & 0 & \cdots & 1 \end{bmatrix} \begin{pmatrix} b_0 \\ b_1 \\ \vdots \\ b_k \end{pmatrix} = \begin{pmatrix} y_1 \\ y_2 - y_1 \\ y_3 - y_2 \\ y_4 - y_3 \\ \vdots \\ y_{k+1} - y_k \end{pmatrix}. \qquad (2.20)$$

This equation demonstrates that if there is any change in value between y_i and y_{i+1}, it can only be attributed to a change in parameter x_i (complicated by random effects if the model is stochastic). The quantity $\Delta y_i = y_{i+1} - y_i$ is an estimate of the effect on y of changing X_i from 0 to 1. It is applicable everywhere if the linear model is appropriate, and for some region around the current sample point otherwise.

As another form of analysis, consider that parameter X_i takes the value 0 in exactly i simulations and the value 1 in $k+1-i$ simulations. Estimates of the average values of Y when parameter X_i takes the values 0 and 1 are given by

$$\hat{y}_{X_i=0} = \frac{1}{i}\sum_{j=1}^{i} y_j \qquad \hat{y}_{X_i=1} = \frac{1}{k+1-i}\sum_{j=i+1}^{k+1} y_j. \qquad (2.21)$$

Equation (2.21) is clearly unbalanced. For instance, if $k = 15$ and $i = 1$, $\hat{y}_{X_i=0}$ is based on one simulation result, whereas $\hat{y}_{X_i=1}$ is calculated from 15 simulation results. A more balanced result can be obtained if the system of equations in (2.19) is expanded to the $2k+1$ equations below, representing a round trip that returns to the starting point. Note that as the first and last equation are the same, only $2k$ of the equations are independent.

$$\begin{bmatrix} 1 & 0 & 0 & 0 & 0 & \cdots & 0 \\ 1 & 1 & 0 & 0 & 0 & \cdots & 0 \\ 1 & 1 & 1 & 0 & 0 & \cdots & 0 \\ \vdots & \vdots & \vdots & \vdots & \vdots & \ddots & \vdots \\ 1 & 1 & 1 & 1 & 1 & \cdots & 1 \\ 1 & 0 & 1 & 1 & 1 & \cdots & 1 \\ 1 & 0 & 0 & 1 & 1 & \cdots & 1 \\ 1 & 0 & 0 & 0 & 1 & \cdots & 1 \\ \vdots & \vdots & \vdots & \vdots & \vdots & \ddots & \vdots \\ 1 & 0 & 0 & 0 & 0 & \cdots & 0 \end{bmatrix} \begin{pmatrix} b_0 \\ b_1 \\ \vdots \\ b_k \end{pmatrix} = \begin{pmatrix} y_1 \\ y_2 \\ y_3 \\ \vdots \\ y_{k+1} \\ y_{k+2} \\ y_{k+3} \\ y_{k+4} \\ \vdots \\ y_1 \end{pmatrix}. \quad (2.22)$$

Equation (2.23) is analogous to Equation (2.20) for this expanded set of equations.

$$\begin{bmatrix} 1 & 0 & 0 & 0 & 0 & \cdots & 0 \\ 0 & 1 & 0 & 0 & 0 & \cdots & 0 \\ 0 & 0 & 1 & 0 & 0 & \cdots & 0 \\ \vdots & \vdots & \vdots & \vdots & \vdots & \ddots & \vdots \\ 0 & 0 & 0 & 0 & 0 & \cdots & 1 \\ 0 & -1 & 0 & 0 & 0 & \cdots & 0 \\ 0 & 0 & -1 & 0 & 0 & \cdots & 0 \\ 0 & 0 & 0 & -1 & 0 & \cdots & 0 \\ \vdots & \vdots & \vdots & \vdots & \vdots & \ddots & \vdots \\ 0 & 0 & 0 & 0 & 0 & \cdots & -1 \end{bmatrix} \begin{pmatrix} b_0 \\ b_1 \\ \vdots \\ b_k \end{pmatrix} = \begin{pmatrix} y_1 \\ y_2 - y_1 \\ y_3 - y_2 \\ \vdots \\ y_{k+1} - y_k \\ y_{k+2} - y_{k+1} \\ y_{k+3} - y_{k+2} \\ y_{k+4} - y_{k+3} \\ \vdots \\ y_1 - y_{2k} \end{pmatrix}. \quad (2.23)$$

This system of equations can yield two estimates of the effect of changing parameter X_i. First, the quantity

$$\Delta_i Y = \frac{1}{2}(y_{i+1} - y_i - y_{k+i+1} + y_{k+i}) \quad (2.24)$$

is a revised estimate of the effect on Y of changing X_i from 0 to 1. It combines the two rows in Equation (2.23) where the only parameter change is in X_i, and it would apply to the entire domain if Y varied linearly with X_i.

The other estimate, $(\hat{y}_{X_i=1} - \hat{y}_{X_i=0})$, is based on Equation (2.21). With the $k-1$ extra data points in Equation (2.22), the number of data points contributing to each of these average quantities is now more balanced, as each uses k data points:

$$\hat{y}_{X_i=0} = \frac{1}{k}\left(\sum_{j=1}^{i} y_j + \sum_{j=k+i+1}^{2k} y_j\right) \quad \hat{y}_{X_i=1} = \frac{1}{k}\sum_{j=i+1}^{i+k} y_j. \qquad (2.25)$$

In the OAT experimental design shown here, each parameter X_i takes only two distinct values. These have been shown as the extreme values, 0 and 1, primarily to simplify the equations. More complicated paths through the sample space would involve much smaller changes to the parameter values, which would allow the estimation of elementary effects, as mentioned in Chapter 1. This approach is described in more detail in Chapter 3.

Another simplification is that the parameters have been changed in the order $X_1, X_2, \ldots X_k$. This choice simplifies Equations (2.17) through to (2.25). However, the parameters could be changed in any order with similar results, though the analyst would need to know the order to be able to compute average values.

With the choices described here, each parameter change corresponds to a move from one corner of the sample hypercube to an adjacent corner, along an edge of the hypercube. If the order of the parameter changes is determined randomly, the path defined by the successive parameter changes is a random walk along the surface of the sample hypercube. In Equation (2.22), the sample point for simulation $l > k$ is diametrically opposite the point for simulation $l - k$. (Such points, placed symmetrically with respect to the centre of the hypercube, are called *mirror points*.) As a result, the design described by Equation (2.22) is a walk along the surface of the sample hypercube from a starting corner to the opposite corner, and then back again following the mirror points of those used on the outward path. If the order of the parameter changes is randomized, the path becomes a random walk to the opposite point, followed by a return using mirror points.

If a function Y follows a linear model as shown in Equation (2.16), the average of two mirror point values is simply the constant $b_0 + \frac{1}{2}\sum_{r=1}^{k} b_r$, which is the value at the centre of the hypercube, where every parameter takes the value 0.5. Averages of all the pairs of mirror points in an OAT design give a set of values that all approximate to $b_0 + \frac{1}{2}\sum_{r=1}^{k} b_r$. The variance of this set is one measure of the nonlinearity of the function.

Values other than 0 and 1 could be used in Equations (2.17) and (2.22). Typically, an analyst would use values symmetrically placed around 0.5, such as 0.2 and 0.8, so that points would be mirror points. The earlier section on sampling values for a single parameter suggested an alternative, but with only two values, the options are quite limited. A later chapter discusses how the number of options can be expanded for this design.

2.4.3 Limits on the Number of Influential Parameters

OAT sampling is inefficient when the number of parameters k is large and only a few of them are influential. Each simulation changes the value of one parameter. If only a few parameters are influential, most of the simulations would be devoted to determining the very small effects of noninfluential parameters. These simulations would be duplicates as far as the values of influential parameters are concerned. Very little new information would be generated.

It is not unusual to have only a few influential parameters. In fact, even when the number of parameters is large, the number of *influential* parameters is *always* small. Why is this the case?

The number of influential parameters in a model is akin to the number of exceptional individuals in a group. For instance, it is common for a school to have a small number of exceptional athletes who can outperform everyone else at running events. Yet in some years there may be no exceptional performances in a particular running event; all the times may be ordinary. Do we then say that every participant is exceptional? No – we reserve that term for situations in which one or a few athletes stand out from the rest.

Consider the scatterplot in Figure 2.4, showing values of Y resulting from a random selection of sampled parameters. Assume Y is a linear combination of several parameters (as in Equation 2.16), and the parameter

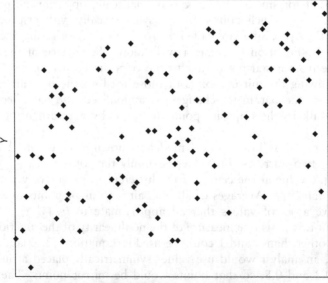

Marginally influential parameter X_r

Figure 2.4 X_r contributes 10% of the variance of Y

SENSITIVITY ANALYSIS OF MULTIPLE PARAMETERS

X_r against which Y is plotted explains 10% of the variance of Y. The figure shows some influence of X_r on Y, but this parameter would be considered only marginally influential.

The linear function in Equation (2.16) has a variance that depends on the sum of squares of the coefficients $b_1, b_2, \ldots b_k$. If the number of influential parameters in a linear model comes to be as large as 10, the least influential parameter cannot contribute more than 10% of the variance, and it can contribute that much only if the influential parameters are all equally influential.

With a nonlinear function, we can use the facts that first, the sensitivity coefficients $S_{ij\ldots l}$ described in Chapter 1 are all nonnegative, and second, they sum to 1. These sensitivity coefficients cannot all be influential, as they are competing with each other to influence the output variable Y. Only a small number can win that competition. Certainly one could not have hundreds of influential parameters.

It is possible that there are *no* influential parameters for sensitivity analysis to reveal. For instance, consider the model

$$Y = \sum_{r=1}^{k} X_r \qquad (2.26)$$

in which all the parameters play exactly the same role. Sensitivity analysis can reveal that many parameters have similar small effects, but cannot give further insight into this model's behaviour. All of the parameters are equally *noninfluential*.

Usually, however, there are a few parameters that stand out from the rest. So with large numbers of parameters, it is desirable to find a more efficient approach than changing one parameter at a time.

2.4.4 Fractional Factorial Sampling

Table 2.1 shows a two-level factorial design for three parameters. These eight simulations contain all possible combinations of low and high values for X_1, X_2 and X_3, where by convention '-1' represents a low value, and '1' represents a high value.[5] These eight points occupy the eight corners of the cube shown on the right, which represents the sample space for

---

[5] For practical reasons it is traditional to use the bounds -1 and 1 for each parameter X_i in a factorial design. The values in a column of the design form a *contrast*, a set of coefficients that tell you how to combine simulation results to estimate the influence of a particular parameter. Values from a factorial design can be scaled linearly to the interval $[0, 1]$, used previously in this chapter for parameter values, by the transformation $X_i' = (X_i + 1)/2$.

three parameters. In general, k parameters would require 2^k simulations to generate all combinations for a factorial design, and these combinations would represent the corners of the corresponding k-dimensional hypercube. One can generate full factorial designs with $s > 2$ levels for each parameter, to determine behaviour at a grid of locations inside the hypercube. Then the number of simulations would be s^k, which is a much larger quantity than 2^k. Because of the explosive growth of the quantity s^k, two levels are typically used for full factorial designs, except when k is very small.

One benefit of using a full factorial design in designing simulations is that the analyst has the data to estimate the mean Y value for each level of every parameter X_r, by averaging over simulations where all other parameters take all possible combinations of high and low values. One disadvantage of using a factorial design is the enormous number of simulations required. Using two levels, 10 parameters would require $2^{10} = 1024$ simulations, and 20 parameters would need more than a million. However, it is possible to select a fraction of these simulations to generate a smaller, feasible design that can still produce useful results.

Consider for example, the design for seven parameters in Table 2.2. The original three columns from Table 2.1 now appear in columns 1, 2 and 4, and the other four columns are obtained from the original three by multiplication as shown in the column headers. Note that exactly half the values in each column are 1, and that half are -1. Observe also that any *two* columns i and j of this design have the property that the four combinations $(1, 1)$, $(1, -1)$, $(-1, 1)$ and $(-1, -1)$ each occur twice. (For instance, in the first two columns of the table, $(1, 1)$ occurs in rows 1 and 5, and $(-1, -1)$ occurs in rows 4 and 8.)

In general, if one starts with a full factorial design on two levels for n parameters, requiring $k = 2^n$ simulations, it can be converted into a fractional factorial (FF) design for $k - 1$ parameters. Tables 2.1 and 2.2

Table 2.1 A two-level full factorial design for three parameters

X_1	X_2	X_3
1	1	1
-1	1	1
1	-1	1
-1	-1	1
1	1	-1
-1	1	-1
1	-1	-1
-1	-1	-1

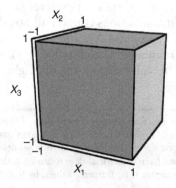

SENSITIVITY ANALYSIS OF MULTIPLE PARAMETERS

Table 2.2 A two-level fractional factorial design for seven parameters

X_1	X_2	$X_3 = X_1X_2$	X_4	$X_5 = X_1X_4$	$X_6 = X_2X_4$	$X_7 = X_1X_2X_4$
1	1	1	1	1	1	1
−1	1	−1	1	−1	1	−1
1	−1	−1	1	1	−1	−1
−1	−1	1	1	−1	−1	1
1	1	1	−1	−1	−1	−1
−1	1	−1	−1	1	−1	1
1	−1	−1	−1	−1	1	1
−1	−1	1	−1	1	1	−1

started with $n = 3$ and produced an FF design with $k = 2^3 = 8$ simulations for $k - 1 = 7$ parameters. Doing this by means of taking products of columns can be an involved and error-prone procedure. There is a much simpler way of generating a design like that of Table 2.2, namely by using Hadamard matrices.

A Hadamard matrix is a matrix of 1's and −1's with the property that it is orthogonal (that is, the product of an $n \times n$ Hadamard matrix and its transpose yields a multiple of the $n \times n$ identity matrix). The smallest Hadamard matrix is H_2:

$$H_2 = \begin{bmatrix} 1 & 1 \\ 1 & -1 \end{bmatrix}. \tag{2.27}$$

When H_2 is multiplied by its transpose (which is simply H_2, since it is symmetric), the result is twice the 2×2 identity matrix, I_2.

$$H_2 H_2^T = \begin{bmatrix} 1 & 1 \\ 1 & -1 \end{bmatrix} \begin{bmatrix} 1 & 1 \\ 1 & -1 \end{bmatrix} = \begin{bmatrix} 2 & 0 \\ 0 & 2 \end{bmatrix} = 2I_2. \tag{2.28}$$

As a generalization of Equation (2.28), $H_n H_n^T = nI_n$ for any $n \times n$ Hadamard matrix H_n. A Hadamard matrix formed in this way makes a good FF design template. It is easy to generate a $2^n \times 2^n$ Hadamard matrix recursively, using this identity:

$$H_{2^n} = \begin{bmatrix} H_{2^{n-1}} & H_{2^{n-1}} \\ H_{2^{n-1}} & -H_{2^{n-1}} \end{bmatrix}. \tag{2.29}$$

Specifically,

$$H_4 = \begin{bmatrix} H_2 & H_2 \\ H_2 & -H_2 \end{bmatrix} = \begin{bmatrix} 1 & 1 & 1 & 1 \\ 1 & -1 & 1 & -1 \\ 1 & 1 & -1 & -1 \\ 1 & -1 & -1 & 1 \end{bmatrix} \tag{2.30}$$

and one more application of the recursion in Equation (2.29) gives H_8, which is the same as Table 2.2, except that H_8 has a column of 1's on the left.

A Hadamard matrix makes a design of *Resolution III*, which in practical terms means that any two columns have an equal number of the combinations $(1, 1), (1, -1), (-1, 1)$ and $(-1, -1)$, as mentioned above. We can construct an FF design matrix with even better properties in the following way. Define M_k for $k = 2^n$ as shown in Equation (2.31).

$$M_k = \left[\frac{H_k}{-H_k} \right]. \tag{2.31}$$

The matrix M_k has k columns and $N = 2k$ rows. It can be used as a design of *Resolution IV*. That means that this design has a much stronger version of the rule given above for any two columns in Table 2.2. Any *three* columns p, q and r of M_k have the property that the eight combinations $(1, 1, 1)$, $(1, 1, -1)$, $(1, -1, 1)$, $(1, -1, -1)$, $(-1, 1, 1)$, $(-1, 1, -1)$, $(-1, -1, 1)$ and $(-1, -1, -1)$ appear equally often in those columns. To put it another way, the variables X_p, X_q and X_r define a cube-shaped sample space like the one shown beside Table 2.1. Each row in M_k then allocates values to X_p, X_q and X_r so that they occupy one of the corners of the cube. In a design constructed from M_k, an equal number of simulations are allocated to each corner of the design cube, *no matter which three parameters we consider*. This balance in distribution is very helpful in determining which parameter and which combinations of parameters working together influence the value of Y.

Utilizing M_k produces a design with $k = 2^n$ columns and $N = 2k = 2^{n+1}$ simulations. This is the same size as the OAT design in Equation (2.22) if k is a power of 2. If k is not a power of 2, we can treat it as one by adding to the set of parameters enough 'dummy' or spare parameters[6] to make the total equal the next higher power of 2.

In the case of FF analysis, a parameter's contrast (the combination of -1's and 1's that appear in the design) also represents the values of several different interactions among two or more other parameters. For instance, in Table 2.2 X_3 was explicitly defined to equal the interaction $X_1 \times X_2$. In a design of Resolution IV, no parameter has the same contrast as any other, nor as any two-factor interaction, but each parameter does have the same contrast as some products of three or more other parameters. Therefore, if a dummy parameter cannot be causing the estimated effect, it is likely that an interaction among influential parameters is causing it.

---

[6] They are called dummy parameters because they do not appear in the model, and therefore cannot influence the output of the model. Dummy parameters act as sentinels. If sensitivity analysis purports to show that a dummy parameter is influential, the analyst knows that something else is at work.

SENSITIVITY ANALYSIS OF MULTIPLE PARAMETERS

One advantage of using the OAT design in Equation (2.22) is that each simulation changes only one parameter, and so, in a deterministic model, the analyst can determine exactly what effect is caused by changing the parameter.[7] One disadvantage is that parameter values for different parameters may be highly correlated. In Equation (2.22), consecutive parameters like X_4 and X_5 differ in value in only two equations out of the total. The average values estimated in Equation (2.25) will therefore be quite similar.

By contrast, designs based on the matrix in Equation (2.31) require substantial computational effort to determine the influence of each parameter, but the parameters themselves have independent values that are not correlated.

One traditional measure of the effect of a parameter in FF analysis is $\mathrm{ME}_r(Y)$, the *main effect* of parameter X_r on Y. This quantity is obtained by taking half the difference of average Y values for the two values of the parameter. As the following equation shows, this definition is proportional to the dot product between the contrast for the parameter and the vector of simulation results.

$$\begin{aligned}
\mathrm{ME}_r(Y) &= \frac{1}{2}\left(\frac{1}{k}\sum_{x_{jr}=1} y_j - \frac{1}{k}\sum_{x_{jr}=-1} y_j\right) \\
&= \frac{1}{2}\left(\frac{1}{k}\sum_{x_{jr}=1} x_{jr} y_j + \frac{1}{k}\sum_{x_{jr}=-1} x_{jr} y_j\right) \\
&= \frac{1}{2k}\sum_{j=1}^{2k} x_{jr} y_j.
\end{aligned} \quad (2.32)$$

It can be shown for a two-level FF design that the variance in the results explained by parameter X_r alone is just

$$\tilde{V}_{Yr} = [\mathrm{ME}_r(Y)]^2. \quad (2.33)$$

A main effect could be estimated for OAT sampling using the estimated means in Equation (2.25). However, the quantity $\Delta_i Y$ from Equation (2.24) typically gives a better estimate of the effect of changing X_i because of the correlations among parameter values that affect the values in Equation (2.25). In any case, the variance property in Equation (2.33) would not apply to OAT sampling. It stems from the orthogonality of the Hadamard matrix used in computing M_k.

---

[7] Of course, this information about changing one parameter applies only at the current point in the sample space. Changing that parameter with some other combination of values for other parameters might have quite a different effect in a nonlinear model.

Equation (2.33) creates a variance partitioning that is very useful in making a quick search for influential parameters, when influence is defined in terms of variance. Of course, the fact that the variance due to X_r is based on the main effect, a linear estimator, emphasizes that variance due to nonlinear effects is not accessible with this (or any other) two-level design.

2.4.5 Latin Hypercube Sampling

A two-level FF design gives no insight into the variation of an output Y as a result of small changes in an influential parameter X_r inside its domain. Section 2.3 discussed how to sample values of a single parameter considered alone. Sometimes a single influential parameter is all we need to study in sensitivity analysis, as all other parameters may have little influence. Just as a two-level FF design is a small fraction of two-level full factorial design, so we can define an s-level FF design that is a small fraction of an s-level full factorial design. How do we select such a fraction with useful properties? The following common method of selection is called Latin hypercube (LH) sampling.

The key objective in LH sampling is to ensure that each parameter is individually stratified over $s > 2$ levels, and that each level contains the same number of points. These conditions require that s divides into N, the sample size (some definitions require that $s = N$). Table 2.3 shows 9 parameters and 6 levels, with each parameter having 2 simulations at each level, for a total of 12 simulations.

Table 2.3 An LH design with 12 simulations for 9 parameters on 6 levels

X_1	X_2	X_3	X_4	X_5	X_6	X_7	X_8	X_9
0	4	5	0	5	4	0	2	0
4	4	1	0	4	3	5	1	4
2	2	4	5	1	0	2	0	0
5	0	5	3	0	5	3	3	5
3	2	3	2	2	5	3	2	3
2	5	2	4	2	1	4	1	1
1	3	3	3	3	4	1	5	5
5	1	0	2	1	3	2	0	4
1	5	4	1	4	2	5	4	2
0	1	1	1	5	1	1	3	1
3	0	2	5	3	0	0	4	2
4	3	0	4	0	2	4	5	3

A randomized design like that of Table 2.3 is easy to generate. Simply store the quantities 0, 0, 1, 1, 2, 2, 3, 3, 4, 4, 5, 5 in each column of an array, and then randomize each column separately.[8] The result is a design in coded form, using integer values 0 to $s-1$, rather than parameter values from the interval [0, 1].

Table 2.4 is also a 12×9 LH design on six levels, but it has a feature that Table 2.3 does not. As indicated by the horizontal line separating top and bottom, Table 2.4 consists of two LH designs, each with six simulations. That means that in each column, the numbers 0 to 5 appear above and below the line.

What is the practical difference between these designs? When Table 2.3 is used to generate parameter values for simulations, an analyst can estimate the mean value of output Y with the sample mean (Equation (2.5)), but it is difficult to assess how good the estimate is. Because of the stratification, it is likely that the estimate is better than what a random sample of the same size would yield, but the traditional uncertainty estimate in Equation (2.5) does not apply.

By contrast, the repeated design in Table 2.4 permits its own uncertainty estimates. Any statistic that can be calculated separately for the top and the bottom of the table, such as a sample mean for Y, now has two independent

Table 2.4 A doubled LH design with 12 simulations, 9 parameters, 6 levels

X_1	X_2	X_3	X_4	X_5	X_6	X_7	X_8	X_9
0	4	5	0	5	4	0	2	0
4	3	1	1	4	3	5	1	4
2	2	4	5	1	0	2	0	2
5	0	0	3	0	5	3	3	5
3	1	3	2	2	2	1	4	3
1	5	2	4	3	1	4	5	1
2	3	3	3	3	4	1	5	5
5	1	0	2	1	3	2	0	4
1	5	4	1	4	2	5	4	2
0	2	1	0	5	1	3	3	1
3	0	2	5	2	0	0	1	0
4	4	5	4	0	5	4	2	3

---

[8] An array a of length N can be randomized using a function $random(N)$ that generates a pseudo-random integer value from 0 to $N-1$, and a method $swap$ that interchanges two entries of the array:
```
for ( int j = 0 ; j < N; j++ )
swap(a, j, random(N)) ;
```

estimates. The uncertainty in these estimates can be obtained directly from the standard deviation of the two. That is,

$$\bar{Y}^{[1]} = \frac{1}{6}\sum_{i=1}^{6} y_i$$

$$\bar{Y}^{[2]} = \frac{1}{6}\sum_{i=7}^{12} y_i$$

$$\bar{Y} = \frac{1}{2}\left(\bar{Y}^{[1]} + \bar{Y}^{[2]}\right) \tag{2.34}$$

$$s_{\bar{Y}} = \sqrt{\left(\bar{Y}^{[1]} - \bar{Y}\right)^2 + \left(\bar{Y}^{[2]} - \bar{Y}\right)^2} = \frac{1}{\sqrt{2}}\left|\bar{Y}^{[1]} - \bar{Y}^{[2]}\right|. \tag{2.35}$$

Now examine Table 2.5. It is the same as Table 2.4, except that some randomly selected levels have been marked with a prime. Every column now has 12 levels: 0 and 0′, 1 and 1′, …5 and 5′. If we replace these symbols by 0, 1, 2, …11 respectively, we have an LH design with 12 levels. But because of the way it was constructed, the bottom and top halves are also LH designs on six levels. By constructing a two-level design this way, we can use the uncertainty estimate of Equation (2.35), while retaining the added benefit of using an LH sample on 12 levels. The uncertainty estimate will probably be an overestimate as a result.

The preceding discussion has addressed the issue of generating integer levels. To convert the integer levels to parameter values, the analyst can choose from several options. The design's integer value can be interpreted as identifying one subinterval out of s equal-length subintervals. The actual value to be used within that subinterval can be randomly generated from

Table 2.5 An LH design with 12 simulations, 9 parameters, 12 levels

X_1	X_2	X_3	X_4	X_5	X_6	X_7	X_8	X_9
0′	4′	5	0	5′	4	0	2	0′
4′	3′	1	1′	4	3	5	1	4′
2′	2	4′	5	1	0	2′	0′	2
5	0′	0	3	0	5	3′	3′	5′
3	1	3′	2′	2	2′	1	4	3
1	5	2	4	3	1	4′	5′	1′
2	3	3	3′	3′	4′	1′	5	5
5′	1′	0′	2	1′	3′	2	0	4
1′	5′	4	1	4′	2	5′	4′	2′
0	2′	1′	0′	5	1′	3	3	1
3′	0	2′	5′	2′	0′	0′	1′	0
4	4	5′	4′	0′	5′	4	2′	3′

SENSITIVITY ANALYSIS OF MULTIPLE PARAMETERS

a uniform distribution, or it can be chosen at a fixed location within the interval (e.g. the midpoint). These are the same options illustrated in Figure 2.2 for a single parameter. One can even divide the interval [0, 1] into $s-1$ (rather than s) subintervals, and interpret the numbers in the table as the s endpoints of these subintervals. This approach would lead to a piecewise linear fit of Y to each influential parameter, independently. Higher-order fits could be attempted to the mesh of points defined by interval endpoints for two or more influential parameters.

LH designs have attractive properties because values sampled for each parameter are stratified. In particular, one can use an LH sample mean as an estimate of a population mean. This estimate converges to the true mean more rapidly with increasing N than does a sample mean from a random sample. If Y is monotonic in each of the parameters X_r, a sample mean will have less uncertainty with LH sampling than a random sample mean at any sample size.

When the number of simulations N is much larger than the number of parameters k, a randomized LH design can be very effective in examining the influence of each parameter, through scatterplots and regression analysis. However, when $N \leqslant k$, the effects of different parameters cannot all be distinguished, because there are not enough data points to give independent estimates for each one. In this case the columns of the design matrix display excessive correlation. Excessive correlation between pairs of columns can persist even when N is somewhat greater than k. Then LH designs need extra structure, and that structure can be provided in various ways.

For example, orthogonal arrays are arrays of integers with properties similar in nature to those of a Hadamard matrix, but which can be much more complex. Orthogonal arrays can be used to generate LH samples where there is some degree of control over the similarities between columns for different parameters, to maximize the information available from a design of a particular size. Table 2.6 has nine simulations for four parameters operating at three levels. It has the additional balance property that each

Table 2.6 A 9×4 LH design on three levels based on an orthogonal array

X_1	X_2	X_3	X_4
0	0	0	0
0	1	1	2
0	2	2	1
1	0	1	1
1	1	2	0
1	2	0	2
2	0	2	2
2	1	0	1
2	2	1	0

pair of columns contains all the nine combinations (0, 0), (0, 1), (0, 2), ... (2, 1), (2, 2). It is not easy to create a new orthogonal array, but they have been tabulated for many different sizes (Colbourn and Dinitz, 1996).

The next section describes a different way of adding structure to an LH sample using a two-level FF design.

2.4.6 Multivariate Stratified Sampling

The principle behind stratified sampling with a single variate was to partition the sample space into nonoverlapping regions and to guarantee sampling from each region. With a single parameter, the regions are contiguous intervals. The purpose for stratified sampling is to ensure that all parts of the sample space are represented, for improved (i.e. less uncertain) mean and variance estimates. This goal is just as important in multivariate sampling, but harder to achieve.

Suppose one pursues the goal by bisecting the domain for every parameter and then taking at least one sample point from each 'corner' of the sample space. One parameter has two 'corners', the regions $[0, 0.5)$ and $[0.5, 1]$. (The brackets indicate that the point 0.5 is in the second region only.) Two parameters, X_p and X_q, have four 'corners', the quadrants of the $X_p X_q$ plane. Three parameters X_p, X_q and X_r have eight 'corners', the octants of the $X_p X_q X_r$ space. In general, k parameters have a sample space that is a k-dimensional hypercube, with 2^k 'corners'. The first three cases are illustrated in Figure 2.5.

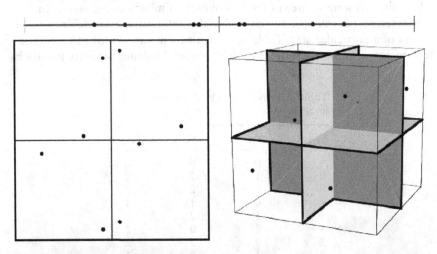

Figure 2.5 Bisecting the domain of each parameter for one [top], two [left] or three [right] parameters

SENSITIVITY ANALYSIS OF MULTIPLE PARAMETERS

It would be desirable to have at least one point in each corner of the sample space, but for large numbers of parameters, it is usually infeasible. Attempting to do so would be equivalent to generating a two-level full factorial design, requiring 1024 simulations for 10 parameters, over a million for 20 parameters, and 2^k simulations for k parameters.

However, the FF design approach partially solves this problem. Earlier we used the interpretation that a value of '-1' in the design matrix M_k should correspond to the single parameter value 0 for X_r, and that '1' in the design matrix should correspond to a parameter value of 1. Suppose instead that we use the interpretation that a '-1' in an FF design means sampling from the interval $[0, 0.5)$ and a '1' means to sample from the interval $[0.5, 1)$. Then the properties of a Resolution IV design will mean that there will be an equal number of points in each of the octants of the sample space *for any three parameters* viewed in isolation. Figure 2.5 would appear approximately as shown for any selection of parameters. This balance helps to expose dependencies of Y on interactions among parameters.

Figure 2.6 shows a plot of a function $Y = \left(\sum_{i=1}^{3} X_r^2\right)^{1/2}$ against X_r for the two different interpretations. In the left panel, all parameter values are either 0 or 1, and very little information about the dependence of Y on X_r is evident. Many of the points overlap since there are only four possible values for Y, corresponding to zero, one, two or three parameters with a value of 1. In the right panel parameter values span the domain of each parameter and greater insight can be gained into the effect of an influential parameter across its domain.

Unfortunately, there is a cost attached to this greater insight. Equation (2.33) (using a main effect to estimate the variance explained by parameter X_r) no longer applies, except as a rough approximation. That equation was based on the assumption that variation in Y due to

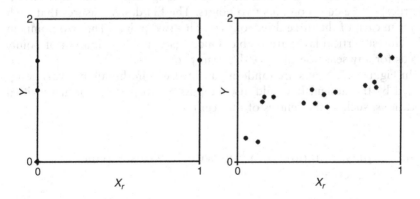

Figure 2.6 Different interpretations of an FF design: [left] '-1' maps to 0, '1' maps to 1; [right] '-1' maps to the left half of the domain, and '1' maps to the right half. Both plots contain 16 points

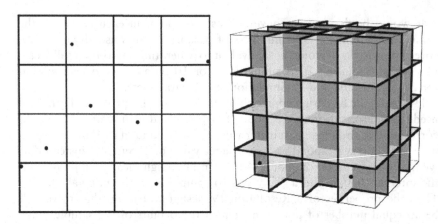

Figure 2.7 Combined fractional factorial – Latin hypercube design

changes in X_r could only occur as a result of the difference between low and high extreme values of X_r. If we let X_r vary within the upper or lower interval, those small changes could also affect the value of Y, and so should contribute to the variance of Y attributable to X_r.

It is possible to combine two-level FF and LH sampling, to secure the advantages of both. Figure 2.7 shows what happens in two and three dimensions when a two-level FF design is combined with stratified sampling inside each of the main two levels. In the pane on the left side, the two-level FF design ensures that each of the four quadrants contains the same number of points (two out of the eight). The LH sample ensures that each of the four columns contains two points, and similarly for the four rows. This behaviour would be observed for any pair of two parameters in the design. In the pane on the right side, the FF design ensures that each octant (corner) of the cube contains two points. The LH design ensures that each layer in each of the three directions contains two points. The two points in the closest vertical layer are visible. Once again, this distribution of points holds for any selection of three different parameters.

In Figure 2.7, points are randomly located within the subintervals designated by the design. It would also be possible to place them at specified locations, such as the centres of the regions.

2.4.7 Quasi-random Sampling with Low-discrepancy Sequences

As experimental designs become more elaborate, more work must be devoted to the theory, implementation and selection of designs. How does one know that a combined FF–LH design with 128 parameters, 256

SENSITIVITY ANALYSIS OF MULTIPLE PARAMETERS

simulations and 64 LH strata was implemented correctly? How can it be expanded to deal with more parameters? What if this design does not yield adequate results – can one somehow make it part of a larger design with more simulations? This section addresses the possibility of decreasing sample complexity by using quasi-random sampling in much the same way one would use pseudo-random sampling as discussed in Section 2.4.1.

Using a pseudo-random generator for simple Monte Carlo sampling is much simpler than using a complex design: as many values as necessary can be generated, and if more parameters or more simulations are desired, it is a simple matter to generate more.

Unfortunately, as shown earlier, samples generated randomly tend to have clusters and gaps. Where a cluster occurs, function values in that vicinity are overemphasized in statistical analysis. Where a gap arises, function values within that gap are not sampled for statistical analysis. The net effect is that mean values estimated with random samples have an uncertainty that diminishes slowly as $1/\sqrt{N}$ (see Equation 2.5). To reduce an estimated uncertainty by a factor of 10, the analyst must increase N by a factor of $10^2 = 100$.

A mathematical measure called *discrepancy* characterizes the lumpiness of a sequence of points in a multidimensional space. Smaller discrepancy values are better for sensitivity analysis (the distribution is less lumpy). Figure 2.8 illustrates the concept in two dimensions. Each small square in the figure occupies $1/4 \times 1/4 = 1/16$ of the area of the unit square. Since there are 20 randomly located points in the figure, each small square should contain 1 or 2 of them, if the points are evenly distributed. However, one of the small squares contains 5 points, and the other none. The discrepancy of a sequence of points is the maximum absolute difference over a specified set of regions between the area fraction and the point fraction. When the regions are squares within the unit square, the discrepancy in Figure 2.8 is at least $5/20 - 1/16 = 3/16$.

Random sequences of k-dimensional points have a relatively high discrepancy, as shown here. But there are infinite sequences of k-dimensional points that behave much better with respect to this measure. They are called *low-discrepancy sequences*. They have the property that as the sequence length N gets very large, the discrepancy shrinks at the theoretically optimal rate. As a result, an estimated mean for a function $Y(X_1, X_2, X_3, \ldots X_k)$ evaluated on points $\{X_{i1}, X_{i2}, \ldots X_{ik}\}_{i=1,N}$ from such a sequence will converge much more quickly than would an estimated mean based on the same number of random points.[9]

---

[9] How quickly? 'For Monte Carlo integration of a smooth function in n dimensions, the answer is that the fractional error will decrease with N, the number of samples, as $(\ln N)^n/N$, i.e. almost as fast as $1/N$' (Press *et al.*, 1997). This comment was made about the Sobol' sequence, but all low-discrepancy sequences have the same asymptotic performance.

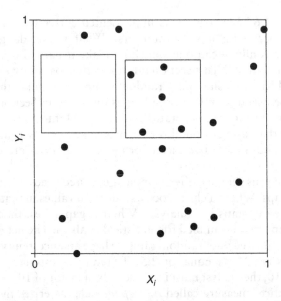

Figure 2.8 Discrepancy is the maximum absolute difference between the fraction of the area a square occupies and the fraction of the points it contains. (Each small square is 1/4 of the width of the unit square.)

Samples made from a finite subset of such sequences are called *quasi-random* samples. These samples are not random, in the sense of being completely unpredictable. In fact, to maintain an even spread of points, an algorithm that generates low-discrepancy sequences must somehow bias the selection of new points to keep them away from the points already present. But they are like random points in the sense that they are uniformly distributed across the entire sample space.

In 'small' quasi-random samples, the effects of low-discrepancy may or may not be evident, as it is an asymptotic property that primarily comes into play as the sample size N gets very large. Whether a sample is large enough to display a low discrepancy depends in part on the number of parameters. When the number of parameters is large, a quasi-random sample will need to be large as well in order for the low discrepancy to become evident.

The Halton sequence is a well-known low-discrepancy sequence that is easy to generate. Table 2.7 shows how to generate values for X_1, X_2, X_3, ... X_{100} for the Halton sequence. On the left side of the table, successive position indices of the points are listed as ordinary numbers (base 10), and in bases 2, 3, 5, ... 541.[10] These bases are the 1st, 2nd, 3rd and 100th

---

[10] The letters *A*, *B*, *C*, etc. are used as digits for base 541 because there are not enough digits in our number system to properly represent base-541 numbers.

Table 2.7 Generating coordinates of points in a Halton sequence using the radical inverse transform

Index in base...				Parameter value			
10	2	3	5	...541	X_1	X_2	X_3	... X_{100}
1	1	1	1	1	$0.1_2 = 0.5$	$0.1_3 = 0.333$	$0.1_5 = 0.2$	$0.1_{541} = 0.002$
2	10	2	2	2	$0.01_2 = 0.25$	$0.2_3 = 0.667$	$0.2_5 = 0.4$	$0.2_{541} = 0.004$
3	11	10	3	3	$0.11_2 = 0.75$	$0.01_3 = 0.111$	$0.3_5 = 0.6$	$0.3_{541} = 0.006$
4	100	11	4	4	$0.001_2 = 0.125$	$0.11_3 = 0.444$	$0.4_5 = 0.8$	$0.4_{541} = 0.007$
5	101	12	10	5	$0.101_2 = 0.625$	$0.21_3 = 0.778$	$0.01_5 = 0.04$	$0.5_{541} = 0.009$
6	110	20	11	6	$0.011_2 = 0.375$	$0.02_3 = 0.222$	$0.11_5 = 0.24$	$0.6_{541} = 0.011$
7	111	21	12	7	$0.111_2 = 0.875$	$0.12_3 = 0.556$	$0.21_5 = 0.44$	$0.7_{541} = 0.013$
8	1000	22	13	8	$0.0001_2 = 0.062$	$0.22_3 = 0.889$	$0.31_5 = 0.64$	$0.8_{541} = 0.015$
9	1001	100	14	9	$0.1001_2 = 0.562$	$0.001_3 = 0.037$	$0.41_5 = 0.84$	$0.9_{541} = 0.017$
10	1010	101	20	A	$0.0101_2 = 0.312$	$0.101_3 = 0.370$	$0.02_5 = 0.08$	$0.A_{541} = 0.018$
11	1011	102	21	B	$0.1101_2 = 0.812$	$0.201_3 = 0.704$	$0.12_5 = 0.28$	$0.B_{541} = 0.020$
12	1100	110	22	C	$0.0011_2 = 0.188$	$0.011_3 = 0.148$	$0.22_5 = 0.48$	$0.C_{541} = 0.022$

largest prime numbers, and in general, X_r is based on the rth largest prime number.

On the right side of the table, the digit sequences of the indices are *reversed*, and placed after a decimal point. This operation is called the *radical inverse transform*. The fractions so generated are then converted back to base 10 decimal fractions for comparison.

Figure 2.9 shows scatterplots of the first three parameters shown in Table 2.7 plotted against each other in pairs. The points look fairly evenly distributed, especially in the lower row where 1000 points are shown in each plot. However, there are also visible patterns in the dots (e.g. diagonal lines of dots). All low-discrepancy sequences have an even distribution, by definition, if the sample is large enough. Different sequences will display different patterns for small numbers of points.

Figure 2.10 shows scatterplots of X_1 against X_{100} with 100 and 1000 points. It is clear from this figure that more points are needed to hide artefacts of the way the points were generated.

One simple way to test a quasi-random sample with N points before using it in sensitivity analysis is to average each variable independently. That is,

$$\int_0^1 X_j dX_j = 0.5 \cong (1/N) \sum_{i=1}^{N} x_{ij}. \qquad (2.36)$$

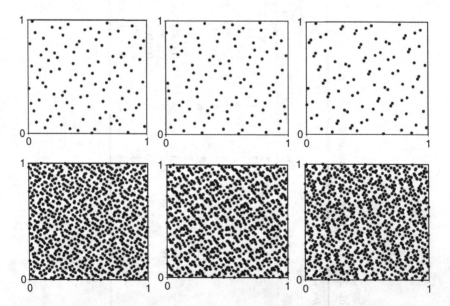

Figure 2.9 Parameters of Table 2.7 plotted against each other in pairs. Left to right: X_2 vs X_1, X_3 vs X_1, X_3 vs X_2. Top row: 100 points; bottom row: 1000 points

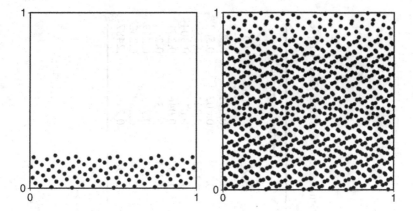

Figure 2.10 X_{100} vs X_1 from the Halton sequence (left: 100 points, right: 1000 points)

This fit should be very close, because of the quasi-random property. Similarly, it is a good idea to check at least some cross-products:

$$\int_0^1 (X_j X_k) dX_j dX_k = 0.25 \cong (1/N) \sum_{i=1}^{N} x_{ij} x_{ik}. \tag{2.37}$$

If the quasi-random sample cannot integrate these functions quite closely, it will probably not perform well in other sensitivity analysis tasks.

Table 2.8 shows how these quantities converge for the variables in Table 2.7, and for a couple of randomly generated variates, U and V. For $N = 10\,000$, parameters X_1, X_2 and X_3 show tighter convergence than would be expected from random variables. However, X_{100} still shows worse convergence than the random variates. There is no apparent advantage at this sample size in using X_{100} in preference to a random variate.

Halton points are not the only quasi-random sequence described in the literature. One of the exercises at the end of this chapter challenges you to find other sequences, of which there are several. One that has been used in sensitivity analysis is the Sobol' LP_τ sequence, for which Fortran code is given in Press et $al.$ (1997). That code supports up to 51 dimensions (i.e. $k = 51$), and up to about 1 billion values (i.e. $N = 2^{30}$). Code that can handle larger problems is also available. Like the Halton sequence, more points are required for good behaviour in higher dimensions. An expedient to get the best possible results for a given N is to renumber the parameters so that those with lower numbers (e.g. X_1, X_2, X_3) are the influential ones, if known. Another approach that has been explored is to randomize the entries for each parameter separately, to reduce correlations among parameters.

Table 2.8 Sample averages of single parameters and two-parameter products for different quasi-random sample sizes (sample averages should converge to 0.5, and product averages to 0.25)

Sample averages							Product averages		
N	X_1	X_2	X_3	X_{100}	U	V	X_1X_2	X_1X_{100}	UV
1	0.500	0.333	0.200	0.002	0.738	0.135	0.167	0.001	0.099
3	0.500	0.370	0.400	0.004	0.831	0.644	0.139	0.002	0.562
10	0.444	0.441	0.428	0.010	0.660	0.521	0.172	0.004	0.354
30	0.484	0.472	0.470	0.029	0.563	0.522	0.223	0.014	0.304
100	0.490	0.489	0.492	0.093	0.542	0.518	0.237	0.046	0.291
300	0.496	0.495	0.495	0.278	0.524	0.501	0.245	0.138	0.260
1000	0.499	0.498	0.499	0.465	0.510	0.511	0.248	0.232	0.259
3000	0.4995	0.4994	0.4997	0.477	0.506	0.502	0.2494	0.238	0.252
10000	0.4998	0.4998	0.4999	0.492	0.5003	0.4976	0.2498	0.246	0.2487

Quasi-random samples can be analysed just as any empirical data set would be. One can calculate sample means and variances using the usual formulas. An output variable Y can be plotted against an input parameter X_i to look for patterns. Two differences arise. First, traditional uncertainty estimates do *not* apply. For instance, an estimate $\hat{Y}_N$ as calculated in Equation (2.4) would be expected to converge more rapidly than one based on random data as N increases, with the advantage growing for larger N. However, there is no obvious way of determining the uncertainty. The second difference is that small samples might not work well at all, depending on how the points are generated.

2.5 GROUP SAMPLING

WHAT CAN BE DONE IF THE REQUIRED NUMBER OF SIMULATIONS IS NOT AFFORDABLE? IS THERE ANY WAY TO STUDY ONLY THE INFLUENTIAL PARAMETERS, IF THEY ARE NOT KNOWN IN ADVANCE?

Designs considered so far require comparable numbers of simulations. For k parameters, an analyst should expect to use $N = 2k$ simulations[11] or more. An OAT design needs that many to estimate the effect of changing each parameter, and an FF design of that size allows for estimation of main effects for each parameter. While LH samples can be of any size, strategies to reduce the correlations of different parameters in an LH design (such as combining FF with LH) tend to require N or more simulations.

This restriction on the minimum number of simulations could have an adverse effect on a modelling project. A model will generally become more complicated (i.e. acquire more parameters) over time as researchers add more features and elaborate on originally simple submodels. The natural evolution of a model could be curtailed if sensitivity analysis requirements make the simulation requirements with more parameters prohibitively expensive. Researchers need to find a way of reducing simulation requirements for complex models.

One way to reduce the number of simulations required is to apply *group sampling*. It allows the analyst to generate smaller designs that can still isolate influential parameters and their effects. It is even possible to obtain sensitivity analysis information from *supersaturated designs*, where the number of simulations, N, is less than the number of parameters, k.

---

[11] It should be noted that k does not affect the number of sample points required simply to estimate the mean of Y to a given level of uncertainty. The distribution of Y alone determines how many sample points are needed for an accurate estimate of the mean of Y. The number of parameters k does affect sensitivity analysis, however, because the analyst estimates not just one mean value, but also one or more coefficients for every parameter in the analysis.

How can this be so? Look back at Equation (2.17) in Section 2.4.1. In matrix form, it specifies that $X_{Nk}B_k = Y_k$, which is a system of linear equations with N rows and $k+1$ columns. From linear algebra, we know there is no unique solution to these equations when $N < k+1$. Furthermore, we will often be looking for functional forms more complex than a linear relationship, requiring the estimation of more coefficients. It seems unlikely that there is a mechanism for estimating these functions with too few simulations. It can be done only if we make some judicious assumptions.

The primary assumption (the null hypothesis) is that every parameter X_r has a negligible influence on the output variable Y. We will reject the null hypothesis for a particular parameter only if the data provide strong evidence of such an effect. This approach changes the statistical procedure from one of estimation (e.g. estimating the b_j coefficients in Equation 2.17) to one of statistical testing. We test each parameter for its effect, identify those for which we can reject the null hypothesis with high confidence, and then analyse the effects of only those parameters. Here we are aided by the conclusion in Section 2.4.3 that only about a dozen parameters can be influential compared to the rest. A supersaturated design can be effective if we focus on identifying and characterizing a small number of influential parameters.

Box 2.1 'A Counterfeit Coin Puzzle' shows how clever design can extract the maximum amount of information from a small amount of data, given appropriate assumptions. In this puzzle, 12 coins need to be weighed on a balance scale only three times in order to find the one counterfeit coin and to tell whether it is lighter or heavier than the rest. A straightforward solution in which each coin is weighed against a standard reference coin would require as many as 11 comparisons, so reducing this number to three is a sound achievement. The solution works only if there is exactly one counterfeit coin. Moreover, the design is sequential: what is weighed in the second and third steps depends on the results of the first and second, respectively.

Box 2.1 A Counterfeit Coin Puzzle (Fixx, 1972)

Suppose you are given 12 apparently identical coins, and you are told that one of them is counterfeit. It can only be told apart from the other coins by its weight – it is either lighter or heavier than a real coin. How many weighings with a balance scale would you need to find the fake coin, and determine if it is heavier or lighter than the rest?

GROUP SAMPLING

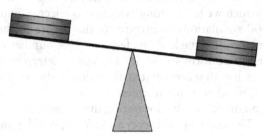

<u>Answer</u>: only three.

1. Number the coins from 1 to 12. First weigh coins 1 to 4 together against coins 5 to 8.

    (i) If the two sides balance, the fake coin is one of 9 to 12. Assume coins 1 to 8 are all good. Weigh 9 to 11 together against 1 to 3.

    (ii) If the two sides still balance, the fake coin must be 12. Weigh it against any other coin to see if it is heavy or light.

    (iii) If the second weighing does not balance, the fake coin is one of 9 to 11, and you know whether it is heavy or light by the way the scale tipped. Suppose it is heavy. Weigh 9 against 10 to see which one is the fake (i.e. the heavy one). If they balance, 11 is the counterfeit.

2. Suppose coins 1 to 4 were heavier than 5 to 8 in the first weighing. Then either the fake is heavy, and one of coins 1 to 4, or light, and one of 5 to 8. Coins 9 to 12 are assumed fair. Weigh coins 1 to 3 and 5 against coins 9 to 11 and 4.

    (i) If the two sides balance, the fake coin is light, and one of 6 to 8. Weigh 6 against 7 to see which one is the fake (i.e. the light one). If they balance, 8 is the counterfeit.

    (ii) If coins 1 to 3 and 5 are heavy in the second weighing, the fake coin is heavy, and must be one of 1 to 3. Weigh 1 against 2 to see which one is the fake (i.e. the heavy one). If they balance, 3 is the counterfeit.

    (iii) If coins 1 to 3 and 5 are light in the second weighing, then either 5 is the fake and it is light, or 4 is the fake on the other side, and it is heavy. Weight 5 against 1 to see which case applies.

Suppose we analyse a model with 1000 parameters, and utilize a sample design with only 400 simulations. By statistical testing, we identify a dozen parameters for which we have strong evidence to reject insignificance. Then we reuse the 400 simulations to investigate the nature of the influence of those 12 parameters. The task seems feasible. The number of simulations actually required to perform sensitivity analysis is determined more by the number of parameters that are influential (which is always a small number) than by the total number of parameters.

What design might be suitable for testing the influences of a large number of parameters? The concept of *group sampling* provides one answer. In group sampling, all the parameters are assigned to a small number of groups. For example, with $k = 1000$ parameters, we could designate 16 groups as follows:

$$G_1 = \{X_1, X_2, X_3, \ldots, X_{64}\}$$
$$G_2 = \{X_{65}, X_{66}, \ldots, X_{128}\}$$
$$\vdots$$
$$G_{16} = \{X_{961}, X_{962}, \ldots, X_{1024}\}$$

(2.38)

where the extra parameters from X_{1001} to X_{1024} would be dummy parameters.

This kind of grouping will be done several times, and each one will have a set of simulations. In each set of simulations, every parameter in a group takes the same sampled parameter values. That is, $x_{i1} = x_{i2} = \ldots = x_{i,64}$ for every simulation i in the set. The group itself acts like a parameter, just as in the coin-weighing example a group of coins would be weighed together. It is not possible to separate the effects of individual parameters, given the results of one set of simulations, just as one weighing tells us little about the weights of individual coins.

Why have 16 groups, rather than 3 or 300? As in the coin-weighing example, the size of the group can be optimized to yield the most information. If the number of groups is very small, influential parameters will be grouped together, and it will be more difficult to separate their effects and to examine their interactions (i.e. nonlinear effects when they are both varied). Accordingly, it is desirable to have more groups than influential parameters. However, if the number of groups is too large, then each set of simulations will be large, and it will not be possible to have more than a few sets of simulations. On the premise that only a dozen parameters can be influential, it makes sense to have 20 or 30 groups. A power of 2, such as 16 or 32, works well because it allows two-level FF designs to be used.

Given 16 groups, an analyst could conduct 32 simulations in an OAT or Resolution IV FF sample, treating the groups as parameters. From the

GROUP SAMPLING

results the analyst could determine which groups were influential. Suppose two parameters are individually influential. If they are in separate groups, then their effects would appear independently as group effects. If they are in the same group, their effects either sum together, or partially cancel each other. Since the number of influential parameters cannot be large, there will typically be some groups that have no influential parameters, and most of the rest will appear in separate groups.

Then the analyst would redo the group analysis using a different allocation of groups, such as

$$G_{12} = \{X_1, X_{17}, X_{33}, \ldots, X_{993}, X_{1009}\}$$
$$G_{22} = \{X_2, X_{18}, \ldots, X_{994}, X_{1010}\}$$
$$\vdots$$
$$G_{16,2} = \{X_{16}, X_{32}, \ldots, X_{1008}, X_{1024}\}. \quad (2.39)$$

Parameters that were grouped together in Equation (2.38) are mostly separated in Equation (2.39). When the analyst carries out another 32-simulation study, each influential parameter will influence a different group from before, in combination with a different collection of other parameters.

The rearrangement from Equation (2.38) to (2.39) is obviously systematic, designed to suggest how different the groupings are. In practice, an analyst can randomly assign parameters to groups. If two influential parameters appear in the same group in the first grouping, there is only a 1/16 chance they will appear together in the second grouping, and in each subsequent grouping. Given enough groupings, the analyst can separate the effects of different influential parameters.

To make this argument quantitative, we denote with M the number of different groupings, each with its set of simulations. If the grouping is done randomly, two influential parameters will be grouped together on average $M/16$ times. For example, if $M = 8$, two influential parameters will share a group, on average, 8/16 times. We would expect to find them together perhaps once, and not likely more than two or three times. With 10 influential parameters, each one will have a probability of more than 50% of not having any other influential parameters in its group each time.

A more difficult problem arises in weeding out noninfluential parameters that masquerade as influential parameters because they share a group several times with influential parameters. For example, suppose that X_1 is influential. The probability that X_1 shares a group t times out of M with a noninfluential parameter X_j is

$$P_{1j}(t, M) = \binom{M}{t} \left(\frac{1}{16}\right)^t \left(\frac{15}{16}\right)^{M-t}. \quad (2.40)$$

If $M = 8$ and $t = 4$, the probability is $P_{1j}(4, 8) = 0.0008$. While this value is small, it may not be small enough. If there are 1000 parameters, there is a significant probability of $1 - (1 - P_{1j}(4, 8))^{999} = 0.56$ that one of the other parameters, probably a noninfluential one, shares a group four times with X_1. And that coincidence could be enough to make that parameter look influential.

If two or more influential parameters have similar effects, the situation becomes even worse. Suppose the three parameters X_1, X_2 and X_3 have significant effects of similar magnitude. Then a noninfluential parameter that shares a group with any one of them several times can appear influential. By analogy with Equation (2.40), the probability of parameter X_j sharing a group with one of these parameters is roughly[12]

$$P_{123j}(t, M) = \binom{M}{t} \left(\frac{3}{16}\right)^t \left(\frac{13}{16}\right)^{M-t} \qquad (2.41)$$

which evaluates to $P_{123j}(4, 8) = 0.038$ for four repetitions in eight groupings. With 1000 parameters, many noninfluential parameters would create a halo around the influential ones, concealing the parameters that are actually influential.

Three simple steps can help to clarify which parameters are influential. Examine Table 2.9, in which the first two steps will be demonstrated. The left

Table 2.9 Analysis of results for three parameters, showing effects of steps to eliminate noninfluential parameters

	Before analysis			Using signs	After X_1 correction		
Grouping	X_1	X_2	X_j	X_j	X_1	X_2	X_j
1	9.5	−5.2	9.5	−9.5	2.0	−5.2	−2.0
2	2.3	2.3	0.2	0.2	−5.2	−5.2	0.2
3	4.7	−4.6	−0.2	0.2	−2.8	−4.6	−0.2
4	9.9	−4.8	2.6	2.6	2.4	−4.8	2.6
5	7.2	−3.9	7.2	−7.2	−0.3	−3.9	−0.3
6	7.7	−0.6	−3.3	−3.3	0.2	−0.6	−3.3
7	9.6	−7.3	9.6	9.6	2.1	−7.3	2.1
8	8.9	−1.3	0.6	−0.6	1.4	−1.3	0.6
Average	7.5	−3.2	3.3	−1.0	0.0	−4.1	0.5

---

[12] This is not an exact formulation because it does not allow for cases where influential parameters are themselves in the same group.

GROUP SAMPLING

side of the table shows the main effects of parameters X_1, X_2 and X_j. Parameters X_1 and X_2 are both influential parameters, with main effects of similar magnitude, but opposite sign. Parameter X_j is noninfluential, but by chance it shares a grouping with X_1 three times out of the eight groupings. As a result, its average effect is larger than that of X_2. The rectangles around some of the numbers represent values that are repeated in the table because two of the tabulated parameters were in the same group. For instance, X_1 and X_j shared a group in Grouping 7, and so both received the same main effect, 9.6.

The first improvement in group sampling is a simple trick that greatly reduces the probability that impostors like X_j will appear. Sample a random sign variable s_{rm} for each parameter X_r in each grouping m from 1 to M. Each sign variable is either 1 or -1, with equal probability. If $s_{rm} = 1$, then parameter X_r follows the group sampling in grouping m. That is, if X_r belongs to group G_1 in the mth grouping, and if the parameter value for group G_1 is 0.69237 in simulation 5, then $x_{5r} = 0.69237$. If $s_{rm} = -1$, on the other hand, parameter X_r takes values complementary to the group value in the mth grouping. In the case above, the value for x_{5r} would not be 0.69237, but rather $1 - 0.69237 = 0.30763$.

The sign variables for each grouping must be stored and used when calculating a sensitivity statistic, such as a main effect. Toggling the sign variable between 1 and -1 toggles the sign of many sensitivity statistics. Since X_1 causes the large main effects in groups where it occurs, these groups always have a positive main effect after corrections are made for the sign variable associated with X_1. The same cannot be said for a noninfluential variable like X_j. It is just as likely to have a negative main effect as a positive one.

The centre section of Table 2.9 shows what happens to X_j if some of the signs are flipped because the sign variable of X_j is different from the sign variable of the controlling parameter in a group. Instead of having three large main effects that add up to make X_j look like an influential parameter, there is some cancellation, and the resulting average effect for X_j is quite small (-1.0). This first step reduces the number of impostors substantially, but alone it is not enough.

The second step occurs at the analysis stage. When the amount of data available is limited, it is important that an analyst perform sensitivity analysis in a stepwise manner. That is, the analyst should identify the most influential parameter, characterize its effect, and try to remove that effect, leaving a set of residuals to analyse, instead of the original Y values. The analyst should then proceed to analyse the residuals in a stepwise manner.

For example, the right side of Table 2.9 shows what happens when the effect of the most influential parameter, X_1, is removed from the analysis. (To make this effect clear, sign variables have not been used on the right-hand side of the table.) The main effect of X_1 is best estimated as 7.5, its average value from all groupings. Assume that the presence of X_1 adds 7.5

to the main effect of any group to which X_1 belongs. Subtract 7.5 from the main effect of all those groups to which X_1 belonged.

The right side of the table shows what happens after the reduction. The remaining average effect of X_1 after this correction is 0.0. The remaining average effect of the other influential parameter, X_2, has strengthened from -3.2 to -4.1. It looks more influential because the cancellation that occurred with X_1 in Grouping 2 has been eliminated. In contrast, the remaining average effect of the noninfluential parameter, X_j, has been reduced from 3.3 to 0.5. The removal of the effect of X_1 eliminates impostors that gained their apparent influence simply by sharing a group with X_1 several times.

These two approaches, sign variables and stepwise analysis, can be used together to increase the capability of the analyst to identify truly influential parameters. There is one more step that can be taken, if the analyst is observant while carrying out the experimental design.

Suppose the analyst carries out one grouping, and looks at the results. They may show one influential group, several influential groups, or the effects of all the groups may be of a similar magnitude. As in the coin puzzle, the analyst has the option of controlling the group assignments in later groupings based on earlier results. Suppose group G_5 out of 16 groups has an effect that is much larger than the effect of any other group. In the next grouping, the analyst could choose to divide the parameters from G_5 into groups G_1 to G_8, while combining all the other parameters into a pool from which to construct G_9 to G_{16}. This would be an effective technique to isolate a single dominant parameter, by allocating it to ever smaller groups.

Alternatively, it may happen that 12 out of 16 groups show effects of similar magnitude. The implication is that there may be a large number of semi-influential parameters, without any really dominant ones. At that point the analyst could choose to allocate 32 groups instead of 16 at the next grouping, to try to separate these semi-influential parameters.

Sequential analysis of this type is very powerful, but not well understood. It illustrates one of many possible lines of inquiry to be followed by future research in sensitivity analysis. The choice of appropriate experimental designs remains an open and fascinating field.

2.6 EXERCISES

1. Suppose a parameter Z is uniformly distributed between 5 and 9 metres.
   (a) What are its pdf $f_Z(z)$ and cdf $F_Z(z)$?
   (b) What transform would convert Z to a parameter X uniformly distributed between 0 and 1? What transform would convert values of X back to Z values?

EXERCISES

(c) What if Z had a triangular distribution instead? That is, suppose the pdf is a straight line between the points $(5m, 0)$ and $(9m, 0.5)$, and 0 elsewhere. Find its pdf, cdf and transforms.

2. Identify a pseudo-random generator in a computer language or software package that you frequently use. For example, Java has `Math.random()` and Excel has `rand()`.

   (a) What is the algorithm by which this generator computes numbers? How long a period does the generator have before it starts to repeat itself?
   (b) How would you regenerate a sequence of numbers so that you could repeat an earlier set of simulations? [Hint: how do you use a random seed with your generator?] How would you generate a new sample that did *not* overlap with the previous one?
   (c) Suppose you want to change both the number of parameters in your model and the number of simulations carried out. How could you arrange to regenerate the same values for parameters that were used before and new values for new parameters and new simulations?

3. Generate a long sequence of random values for the triangular variate Z described in Question 1. Estimate the sample mean and variance using Equation (2.4). How many simulations are needed to estimate these quantities reliably to one decimal place? [Hint: generate multiple sequences of the same length and ensure they have the same mean and variance to one decimal place.]

4. Generate a stratified random sample for the triangular variate Z described in Question 1. Use N points, with two random points sampled from each of $N/2$ equal-length intervals for X. For instance, if $N = 8$, divide the interval from 0 to 1 into four subintervals: $[0, 0.25)$, $[0.25, 0.5)$, $[0.5, 0.75)$ and $[0.75, 1)$ and sample two points in each subinterval.

   (a) How many simulations are needed to estimate the mean value of Z reliably to one decimal place (see previous question)?
   (b) How do the variance estimates $\tilde{V}_Y^a$ and $\tilde{V}_Y^w$ differ, if $Y = Z$?

5. Suppose $v = 3a - 2bc + 4d^2$, where a, b, c and d are parameters, each with values in the domain $[0, 1]$. Write out a list of eight linear equations generated by a randomized return-trip OAT design starting from $\{1, 1, 1, 1\}$.

   (a) Determine the two effects of changes to the values of each parameter, as in Equation (2.24). Which parameter change causes the largest single change in v? Which parameter has the largest difference between its two effects?

(b) What is the average value of v for a at 0 and a at 1, as in Equation (2.25)? What are the average values for the other parameters? Which one has the largest difference between averages?

(c) Which method of assessing the impact of the parameters is most representative in this example?

(d) Calculate the average value for each pair of mirror points in your sample. Can you determine that the model is nonlinear from these averages?

6. Construct a two-level full factorial design for five parameters. How many simulations are there?

   (a) Consider just the fraction of the design consisting of simulations where an odd number of parameters have value -1. How many simulations are there in this fraction?

   (b) How many simulations would be needed for the same five parameters for a full factorial design on five levels?

   (c) In the factorial design on five levels, every parameter has a value that is 0, 1, 2, 3 or 4. How many simulations would there be in the fraction of the design consisting of simulations where the sum of values for all the parameters is a multiple of 5?

7. Use Equation (2.31) to produce a Resolution IV FF design for 12 parameters. [Hint: how many parameters would the design have to handle?] Evaluate Y values for this design, given that $Y = X_1 + 2X_2 + 3X_3 + 4X_7 X_{12}$. Assume each X_r takes values 1 or -1 only.

   (a) Work out the main effects for all the parameters X_1 to X_{12}. How do they relate to the coefficients of the formula for Y?

   (b) Use your design to work out the contrast (i.e. the pattern of 1's and -1's) for $X_7 X_{12}$ and for two other two-factor interactions. Evaluate interaction effects for these interactions by applying these contrasts using Equation (2.32).

   (c) Can you find a three-factor interaction that has the same contrast as X_1?

8. Generate:

   (a) one LH sample with 5 parameters, 12 levels and 12 simulations;

   (b) three LH samples with 5 parameters, 4 levels and 4 simulations each;

   (c) one LH sample with 5 parameters, 12 levels and 12 simulations by combining the three samples in part (b), and splitting levels 0 to 3 into three sub-levels each (e.g. 0, 0' and 0" become new levels 0, 1 and 2);

   (d) an orthogonal array as shown in Table 2.6 that has 16 simulations on 5 parameters at 4 levels, with the property that all the combinations

(0, 0), (0, 1), (0, 2), (0, 3), (1, 0), ... (3, 2), (3, 3) occur for any pair of parameters. [Difficult]

9. Generate a combined FF and LH design with eight simulations and four parameters using the following steps:

   (a) Generate a 4×4 Hadamard matrix, and then replace every -1 by 0 and every 1 by 4. (In general you use 0 and 2^n for a $2^n \times 2^n$ Hadamard matrix.)
   (b) Generate an LH design with four simulations for four variables on four levels. Each column will have a permutation of the numbers 0, 1, 2 and 3. Add this design to the one from step (a) in the following way. If the entry in the modified Hadamard matrix is 4, add the entry in the same location of the LH design to it. If the entry in the modified Hadamard matrix is 0, add to it $3 - w$, where w is the entry in the same location of the LH design. Entries in the new design will now range from 0 to 7, but in each column there will only be four different values.
   (c) Double the number of rows to make a Resolution IV design by adding to the design the complement of every simulation currently there. The complement is found by subtracting each entry in the original row from 7 to get the entry in the new row.

10. Several low-discrepancy sequences are described in the literature. Find papers or books containing algorithms for generating at least three other low-discrepancy sequences beside the Halton points. In each case cite the source, name the sequence (usually named after the person who discovered it), and describe briefly how the numbers are generated.

11. Suppose you have access to a model with the functional form $Y = aX_m - bX_n$ for some unknown a, b, m and n. How many groupings and runs would you need to find the unknowns with high confidence using FF designs, given that the number of parameters is $k = 1000$? What if $k = 10^6$?

2.7 EXERCISE SOLUTIONS

1. Suppose a parameter Z is uniformly distributed between 5 and 9 metres...

   (a)
   $$f_Z(z) = \begin{cases} 0\,\text{m}^{-1} & z < 5\,\text{m} \\ 1/(9\,\text{m} - 5\,\text{m}) & 5\,\text{m} \leq z < 9\,\text{m} \\ 0\,\text{m}^{-1} & 9\,\text{m} \leq z \end{cases} \quad (2.42)$$

   $$F_Z(z) = \begin{cases} 0 & z < 5\,\text{m} \\ (z - 5\,\text{m})/(9\,\text{m} - 5\,\text{m}) & 5\,\text{m} \leq z < 9\,\text{m} \\ 1 & 9\,\text{m} \leq z \end{cases} \quad (2.43)$$

(b) $X = F_Z(Z)$
$Z = 5\,\text{m} + X(9\,\text{m} - 5\,\text{m})$

(c) $$f_Z(z) = \begin{cases} 0\,\text{m}^{-1} & z < 5\,\text{m} \\ 2(z - 5\,\text{m})/(9\,\text{m} - 5\,\text{m})^2 & 5\,\text{m} \leq z < 9\,\text{m} \\ 0\,\text{m}^{-1} & 9\,\text{m} \leq z \end{cases} \quad (2.44)$$

$$F_Z(z) = \begin{cases} 0 & z < 5\,\text{m} \\ ((z - 5\,\text{m})/(9\,\text{m} - 5\,\text{m}))^2 & 5\,\text{m} \leq z < 9\,\text{m} \\ 1 & 9\,\text{m} \leq z \end{cases} \quad (2.45)$$

$X = F_Z(Z)$
$Z = 5\,\text{m} + \sqrt{X}(9\,\text{m} - 5\,\text{m})$.

2. Identify a pseudo-random generator in a computer language or software package that you frequently use. ...Take Excel's `rand()`.

   (a) Prior to 2003, Excel used a poor pseudo-random generator that started to repeat (approximately) after a cycle of about 1 million values. Excel 2003 upgraded `rand` to use a 48-bit pseudo-random generator published by Wichman and Hill (1982). It has a cycle of more than 10^{13} before repetition occurs. The Fortran algorithm is as follows:[13]

   ```
   C IX, IY, IZ SHOULD BE SET TO INTEGER VALUES BETWEEN
   C 1 AND 30000 BEFORE FIRST ENTRY
   IX = MOD(171 * IX, 30269)
   IY = MOD(172 * IY, 30307)
   IZ = MOD(170 * IZ, 30323)
   RANDOM = AMOD(FLOAT(IX) / 30269.0 + FLOAT(IY) / 30307.0 + FLOAT(IZ) / 30323.0, 1.0)
   ```

   (b) According to one website,[14] it is not possible to set the starting seed in Excel 2003's (and Excel 2007's) version of `rand()`. The Visual Basic method `RANDOMIZE()` is supposed to perform this function, but does not. It is possible, however, to generate a fixed sequence in Excel by filling a large number of cells with the formula `=rand()`, and then copying and pasting the values (not the formulas) of these cells into another range of cells of the same size. Thereafter these values will stay fixed. Similarly, it is not possible to guarantee that two different samples will not overlap. However, because of the long cycle, overlap is unlikely unless you use many random numbers. For example, if you use two separate sample of 10 000 numbers, the odds are less than

---

[13] http://support.microsoft.com/kb/828795
[14] http://www.daheiser.info/excel/main/section15.pdf

EXERCISE SOLUTIONS

$10\,000/10^{13} = 10^{-9}$ that the second sequence starts within the range of the first sequence. In a programming language where you can set the random seed, you would regenerate the first sequence and then take the following numbers to form a new sequence, thereby guaranteeing they do not overlap. Some pseudo-random generators (e.g. l'Ecuyer and Andres, 1997) offer the facility of generating independent and nonoverlapping subsequences.

(c) If your simulation generates pseudo-random numbers as it goes along, start from the same seed and you will get the same sequence. If the number of parameters can change between runs, it helps to have a distinct subsequence for each parameter so that the number of parameters will not affect the numbers produced. It is important that if you have a distinct subsequence for each parameter, the subsequence is made much longer than you expect to use, so that you can expand the number of runs later. In Excel, generate a large table of pseudo-random numbers that is made constant by copying and pasting values from a table where the formula = rand() is used. Assign one parameter to each column, and make sure the number of rows is much larger than what you expect to use. If necessary, you can copy more numbers into the bottom of the table to support more simulations.

3. Generate a long sequence of random values for the triangular variate Z described in Question 1. .... no answer provided.
4. Generate a stratified random sample for the triangular variate Z described in Question 1. .... no answer provided.
5. Suppose $v = 3a - 2bc + 4d^2$, where a, b, c and d are parameters, each with values in the domain [0, 1]. Write out a list of eight linear equations generated by a randomized round-trip OAT design starting from $\{1, 1, 1, 1\}$. The coefficients b_j in the following equation are based on Equation (2.16), and are not related to the variable b.

$$\begin{bmatrix} 1 & 1 & 1 & 1 & 1 \\ 1 & 1 & 1 & 0 & 1 \\ 1 & 0 & 1 & 0 & 1 \\ 1 & 0 & 1 & 0 & 0 \\ 1 & 0 & 0 & 0 & 0 \\ 1 & 0 & 0 & 1 & 0 \\ 1 & 1 & 0 & 1 & 0 \\ 1 & 1 & 0 & 1 & 1 \end{bmatrix} \begin{pmatrix} b_0 \\ b_1 \\ b_2 \\ b_3 \\ b_4 \end{pmatrix} = \begin{pmatrix} 5 \\ 7 \\ 4 \\ 0 \\ 0 \\ 0 \\ 3 \\ 7 \end{pmatrix} \quad (2.46)$$

(a)

	a	b	c	d
1st effect	3	0	−2	4
2nd effect	3	−2	0	4
largest				✓
largest difference		✓	✓	
avg v at 0	1	2.5	2.75	0.75
avg v at 1	5.5	4.0	3.75	5.75
difference	4.5	1.5	1.0	5.0
largest difference				✓
mirror point avg	2.5	3.5	3.5	3.5

(2.47)

(b) See table above.
(c) In this case, the single simulation change best measures the coefficient of each parameter. It gets the correct coefficient for a, gets the right coefficient for b and c part of the time, and gets the right coefficient for d, although with two levels it cannot distinguish between a linear effect and a squared effect.
(d) See the table above for averages of each pair of mirror points. Three are the same, but one is different, indicating some nonlinearity. If you plug $a = b = c = d = 0.5$ into the actual formula, you get $v = 2$, which is quite different from these values, indicating a higher degree of nonlinearity.

6. Construct a two-level full factorial design for five parameters. There are 32 simulations, as shown in the following table:

Run	X_1	X_2	X_3	X_4	X_5	Run	X_1	X_2	X_3	X_4	X_5
1	1	1	1	1	1	17	−1	1	1	1	1
2	1	1	1	1	−1	18	−1	1	1	1	−1
3	1	1	1	−1	1	19	−1	1	1	−1	1
4	1	1	1	−1	−1	20	−1	1	1	−1	−1
5	1	1	−1	1	1	21	−1	1	−1	1	1
6	1	1	−1	1	−1	22	−1	1	−1	1	−1
7	1	1	−1	−1	1	23	−1	1	−1	−1	1
8	1	1	−1	−1	−1	24	−1	1	−1	−1	−1
9	1	−1	1	1	1	25	−1	−1	1	1	1
10	1	−1	1	1	−1	26	−1	−1	1	1	−1
11	1	−1	1	−1	1	27	−1	−1	1	−1	1
12	1	−1	1	−1	−1	28	−1	−1	1	−1	−1
13	1	−1	−1	1	1	29	−1	−1	−1	1	1
14	1	−1	−1	1	−1	30	−1	−1	−1	1	−1
15	1	−1	−1	−1	1	31	−1	−1	−1	−1	1
16	1	−1	−1	−1	−1	32	−1	−1	−1	−1	−1

(a) Consider just the fraction of the design consisting of simulations in which an odd number of parameters have value −1. There is one simulation in each row of the table above, for 16 in total.

Run	X_1	X_2	X_3	X_4	X_5
17	−1	1	1	1	1
2	1	1	1	1	−1
3	1	1	1	−1	1
20	−1	1	1	−1	−1
5	1	1	−1	1	1
22	−1	1	−1	1	−1
23	−1	1	−1	−1	1
8	1	1	−1	−1	−1
9	1	−1	1	1	1
26	−1	−1	1	1	−1
27	−1	−1	1	−1	1
12	1	−1	1	−1	−1
29	−1	−1	−1	1	1
14	1	−1	−1	1	−1
15	1	−1	−1	−1	1
32	−1	−1	−1	−1	−1

(b) On five levels, $5^5 = 3125$ simulations would be required.

(c) There would be $5^4 = 625$ simulations in the fraction. Whatever the sum of the first four parameters is, one and only one value for X_5 would make the total sum a multiple of 5.

7. Use Equation (2.31) to produce a Resolution IV FF design for 12 parameters (see next page).

   (a) Main effects: see the last line of the table on the next page. All coefficients of terms linear in the variables are estimated correctly. The effects of the two-way interaction do not affect the main effects.

   (b) Two-way interaction: see table above for the column $X_7 X_{12}$. Other two-way interactions are not shown. The interaction effect of 4 is correct and it is shown in the ME row of the table.

   (c) The product $X_2 X_3 X_4$ has the same contrast (sign structure) as X_1.

8. Generate:

   (a) one LH sample with 5 parameters, 12 levels and 12 simulations (not shown)

Run	X_1	X_2	X_3	X_4	X_5	X_6	X_7	X_8	X_9	X_{10}	X_{11}	X_{12}	X_7X_{12}	Y
1	1	1	1	1	1	1	1	1	1	1	1	1	1	10
2	1	1	1	1	-1	1	1	-1	1	1	-1	-1	-1	-2
3	1	1	1	-1	1	1	-1	1	1	-1	1	-1	1	4
4	1	1	1	-1	-1	1	-1	-1	1	-1	-1	1	-1	-8
5	1	1	-1	1	1	-1	1	1	-1	1	-1	1	1	2
6	1	1	-1	1	-1	-1	1	-1	-1	1	1	-1	-1	-6
7	1	1	-1	-1	1	-1	-1	1	-1	-1	-1	-1	1	0
8	1	1	-1	-1	-1	-1	-1	-1	-1	-1	1	1	-1	-4
9	1	-1	1	1	1	-1	1	1	-1	-1	1	-1	-1	0
10	1	-1	1	1	-1	-1	1	-1	-1	-1	-1	1	1	2
11	1	-1	1	-1	1	-1	-1	1	-1	1	1	1	-1	6
12	1	-1	1	-1	-1	-1	-1	-1	-1	1	-1	-1	1	4
13	1	-1	-1	1	1	1	1	1	1	-1	-1	-1	-1	0
14	1	-1	-1	1	-1	1	1	-1	1	-1	1	1	1	10
15	1	-1	-1	-1	1	1	-1	1	1	1	-1	1	-1	-2
16	1	-1	-1	-1	-1	1	-1	-1	1	1	1	-1	1	4
17	-1	1	1	1	1	1	1	1	-1	-1	-1	-1	-1	-8
18	-1	1	1	1	-1	1	1	-1	-1	-1	1	1	1	-2
19	-1	1	1	-1	1	1	-1	1	-1	1	-1	1	-1	-6
20	-1	1	1	-1	-1	1	-1	-1	-1	1	1	-1	1	4
21	-1	1	-1	1	1	-1	1	1	1	-1	1	-1	-1	0
22	-1	1	-1	1	-1	-1	1	-1	1	-1	-1	1	1	-10
23	-1	1	-1	-1	1	-1	-1	1	1	1	1	1	-1	2
24	-1	1	-1	-1	-1	-1	-1	-1	1	1	-1	-1	1	-4
25	-1	-1	1	1	1	-1	1	1	1	1	-1	1	1	-10
26	-1	-1	1	1	-1	-1	1	-1	1	1	1	-1	-1	2
27	-1	-1	1	-1	1	-1	-1	1	1	-1	-1	-1	1	-4
28	-1	-1	1	-1	-1	-1	-1	-1	1	-1	1	1	-1	8
29	-1	-1	-1	1	1	1	1	1	-1	1	1	1	1	-2
30	-1	-1	-1	1	-1	1	1	-1	-1	1	-1	-1	-1	-6
31	-1	-1	-1	-1	1	1	-1	1	-1	-1	1	-1	1	4
32	-1	-1	-1	-1	-1	1	-1	-1	-1	-1	-1	1	-1	0
ME	1	2	3	0	0	0	0	0	0	0	0	0	4	

(b) three LH samples with 5 parameters, 4 levels and 4 simulations each:

Run	X_1	X_2	X_3	X_4	X_5
1	1	3	0	3	2
2	3	0	1	2	1
3	2	1	3	1	0
4	0	2	2	0	3
5	1	2	1	2	3
6	3	1	3	1	0
7	2	3	2	3	2
8	0	0	0	0	1
9	2	3	0	2	2
10	1	1	1	3	1
11	0	2	3	1	0
12	3	0	2	0	3

(c) a derived LH sample with 5 parameters, 12 levels and 12 simulations:

Run	X_1	X_2	X_3	X_4	X_5
1	4	10	0	9	6
2	9	1	3	8	5
3	8	5	11	4	1
4	2	6	7	2	10
5	5	7	4	6	11
6	10	3	9	5	0
7	6	11	8	10	7
8	1	0	2	1	3
9	7	9	1	7	8
10	3	4	5	11	4
11	0	8	10	3	2
12	11	2	6	0	9

(d) An orthogonal array:

Run	X_1	X_2	X_3	X_4	X_5
1	0	0	0	0	0
2	0	1	1	1	1
3	0	2	2	2	2
4	0	3	3	3	3
5	1	0	1	2	3
6	1	1	0	3	2
7	1	2	3	0	1
8	1	3	2	1	0
9	2	0	2	3	1
10	2	1	3	2	0
11	2	2	0	1	3
12	2	3	1	0	2
13	3	0	3	1	2
14	3	1	2	0	3
15	3	2	1	3	0
16	3	3	0	2	1

9. Generate a combined FF and LH design with 8 simulations and 4 parameters:

Run	X_1	X_2	X_3	X_4
1	4	7	6	5
2	5	1	5	3
3	7	5	0	0
4	6	3	3	6
5	3	0	1	2
6	2	6	2	4
7	0	2	7	7
8	1	4	4	1

10. No answer provided. Hint: look in the ACM journal *Transactions on Mathematical Software*.

11. Use Resolution IV designs with 8 parameters and 16 runs for each grouping. There are only two influential parameters, so the probability is only 1/8 that they will share the same grouping if grouping is done randomly. Suppose in the first set of 16 simulations two groups appear to be influential. In the next grouping, divide the parameters in those groups into four groups, and divide the rest among the other four groups. Again only two groups (at worst) appear to be influential. At the next grouping, divide the parameters in those two into four groups, and all the rest into four groups. And so on. If we start with 1000 parameters, the number of parameters in each influential group will be: $125 \to 63 \to 32 \to 16 \to 8 \to 4 \to 2 \to 1$. A total of eight groupings with 128 simulations would be required. In general, the number of groupings is $1 + \log_2(k/8)$. With $k = 1\,000\,000$, the number of groupings would be about 18, and the number of simulations would be only 288.

3

Elementary Effects Method

3.1 INTRODUCTION

IS THERE A METHOD WHICH CAN COPE WITH COMPUTATIONALLY EXPENSIVE MODELS AND GIVE SENSITIVITY RESULTS WITH A SMALL NUMBER OF MODEL EVALUATIONS?

In the previous chapter we noted that the amount of information revealed via a sensitivity analysis depends heavily on the number of sample points that are simulated and where they are located.

In this chapter we will describe a sensitivity analysis method which is effective in identifying the few important factors in a model that contains many factors, with a relatively small number of sample points properly distributed.

The method is conceptually simple and easy to implement. It belongs to the class of OAT designs described in Chapter 2 but partially overcomes their main limitations. While adhering to the concept of local variation around a base point, this method makes an effort to overcome the limitations of the derivative-based approach by introducing wider ranges of variations for the inputs and averaging a number of local measures so as to remove the dependence on a single sample point.

This method is ideal when the number of input factors is too large to allow the application of computationally expensive variance-based techniques, but at the same time not large enough to demand the use of group techniques such as those described at the end of Chapter 2. With respect to group techniques it has the advantage of examining each factor individually so as

---

Global Sensitivity Analysis. The Primer   A. Saltelli, M. Ratto, T. Andres, F. Campolongo, J. Cariboni, D. Gatelli, M. Saisana and S. Tarantola   © 2008 John Wiley & Sons, Ltd

to avoid the problem of cancellation effects (i.e. two factors, individually influential, may belong to the same group and have effects that partially cancel each other out).

3.2 THE ELEMENTARY EFFECTS METHOD

WHAT IS AN ELEMENTARY EFFECT? HOW CAN I DEFINE A SENSITIVITY MEASURE USING THE ELEMENTARY EFFECTS? HOW CAN I INTERPRET THE SENSITIVITY RESULTS? WHEN CAN I USE THIS METHOD?

The elementary effects (EE) method is a simple but effective way of screening a few important input factors among the many that can be contained in a model. The fundamental idea behind the method is owed to Morris, who introduced the concept of elementary effects in 1991, proposing the construction of two sensitivity measures with the aim of determining which input factors could be considered to have effects which were (a) negligible, (b) linear and additive, or (c) nonlinear or involved in interactions with other factors. An elementary effect is defined as follows. Consider a model with k independent inputs $X_i, i = 1, \ldots, k$, which varies in the k-dimensional unit cube across p selected levels. In other words, the input space is discretized into a p-level grid Ω. For a given value of $\mathbf{X}$, the elementary effect of the ith input factor is defined as

$$EE_i = \frac{[Y(X_1, X_2, \ldots, X_{i-1}, X_i + \Delta, \ldots X_k) - Y(X_1, X_2, \ldots, X_k)]}{\Delta}, \quad (3.1)$$

where p is the number of levels, Δ is a value in $\{1/(p-1), \ldots, 1 - 1/(p-1)\}$, $\mathbf{X} = (X_1, X_2, \ldots X_k)$ is any selected value in Ω such that the transformed point $(\mathbf{X} + \mathbf{e}_i \Delta)$ is still in Ω for each index $i = 1, \ldots, k$, and $\mathbf{e}_i$ is a vector of zeros but with a unit as its ith component.

The distribution of elementary effects associated with the ith input factor is obtained by randomly sampling different $\mathbf{X}$ from Ω, and is denoted by F_i, i.e. $EE_i \sim F_i$. The F_i distribution is finite and, if p is even and Δ is chosen to be equal to $p/(2(p-1))$, the number of elements of F_i is $p^{k-1}[p - \Delta(p-1)]$. Assume, for instance, that $k = 2$, $p = 4$ and $\Delta = 2/3$, for a total number of eight elements for each F_i. The four-level grid in the input space is represented in Figure 3.1. The total number of elementary effects can be counted from the grid by simply keeping in mind that each elementary effect relative to a factor i is computed by using two points whose relative distance in the coordinate X_i is Δ, and zero in any other coordinate.

The sensitivity measures, μ and σ, proposed by Morris, are respectively the estimates of the mean and the standard deviation of the distribution F_i.

THE ELEMENTARY EFFECTS METHOD

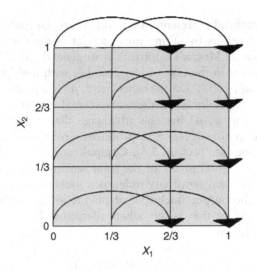

Figure 3.1 Representation of the four-level grid ($p = 4$) in the two-dimensional input space ($k = 2$). The value of Δ is $2/3$. The arrows identify the eight points needed to estimate the elementary effects relative to factor X_1

The mean μ assesses the overall influence of the factor on the output. The standard deviation σ estimates the ensemble of the factor's effects, whether nonlinear and/or due to interactions with other factors. An intuitive explanation of the meaning of σ is the following. Assume that for factor X_i we obtain a high value of σ. The elementary effects relative to this factor thus differ notably from one another, implying that the value of an elementary effect is strongly affected by the choice of the sample point at which it is computed, i.e. by the choice of the other factors' values. By contrast, a low value of σ indicates very similar values among the elementary effects, implying that the effect of X_i is almost independent of the values taken by the other factors. Campolongo *et al.* (2007) proposed replacing the use of the mean μ with μ^*, which is defined as the estimate of the mean of the distribution of the absolute values of the elementary effects that we denote with G_i, i.e. $|EE_i| \sim G_i$.

The use of μ^* is convenient as it solves the problem of type II errors (failing to identify a factor with considerable influence on the model), to which the original measure μ is vulnerable. Type II errors might occur when the distribution F_i contains both positive and negative elements, i.e. when the model is nonmonotonic or has interaction effects. In these cases, some effects may cancel each other out when computing μ, thus producing a low mean value even for an important factor. To avoid such type II errors, Morris (1991) recommended considering the values of μ and σ simultaneously, since a factor with elementary effects of different signs (i.e. that cancel each other out) would have a low value of μ but a considerable value of σ. A graphical representation in the (μ, σ) plane allows for a better interpretation of results by taking into account at the same time the two sensitivity measures. Morris's approach is in general valuable, but it may become problematic in the case of large models with multiple outputs (see Campolongo et al., 2007). On the other hand, μ^* is a practical and concise measure to use, especially when there are several output variables. Moreover, in contrast to μ, μ^* has the advantage that it can be adjusted to work with a group of factors, i.e. to produce an overall sensitivity measure relative to a group (see Section 3.5). Campolongo et al. (2007) have also shown that μ^* is a good proxy of the total sensitivity index S_T discussed in Chapter 4. The total sensitivity index is a measure of the overall effect of a factor on the output (inclusive of interactions) and corresponds to the expected variance that is left when all factors but X_i are fixed, i.e. $E_{X_{\sim i}}\left(V_{X_i}(Y|X_{\sim i})\right)$, and

$$S_{T_i} = \frac{E_{X_{\sim i}}\left(V_{X_i}(Y|X_{\sim i})\right)}{V(Y)}.$$

S_{T_i} is to be used when the goal is that of identifying noninfluential factors in a model (rather than prioritizing the most influential ones). μ^* is an effective substitute for the total index when the computational cost of S_T is unaffordable.

In general, as the estimate of μ comes at no extra computational cost (the same number of model executions is required), we recommend computing all three statistics, μ, σ and μ^*, so as to extract the maximum amount of sensitivity information. For instance, the comparison between μ and μ^* provides information on the signs of the effects that the factor has on the output. If μ and μ^* are both high, it implies not only that the factor has a large effect on the output, but also that the sign of this effect is always the same. If, by contrast, μ is low while μ^* is high, it means that the factor examined has effects of different signs depending on the point of the space at which the effect is computed. The sampling strategy to estimate the three statistics is described in detail in the next section.

3.3 THE SAMPLING STRATEGY AND ITS OPTIMIZATION

HOW CAN I BUILD AN EFFICIENT SAMPLE TO ESTIMATE THE ELEMENTARY EFFECTS?
In order to estimate the sensitivity measures (i.e. the statistics of the distributions F_i and G_i), the design focuses on the problem of sampling a number r of elementary effects from each F_i. As the computation of each elementary effect requires two sample points, the simplest design would

THE SAMPLING STRATEGY AND ITS OPTIMIZATION

require $2r$ sample points for each input, for a total of $2rk$, where k is the number of input factors. Morris (1991) suggested a more efficient design that builds r trajectories of $(k+1)$ points in the input space, each providing k elementary effects, one per input factor, for a total of $r(k+1)$ sample points.

The trajectories are generated in the following manner. A *base* value $\mathbf{x}^*$ for the vector $\mathbf{X}$ is randomly selected in the p-level grid Ω. $\mathbf{x}^*$ is not part of the trajectory but is used to generate all the trajectory points, which are obtained from $\mathbf{x}^*$ by increasing one or more of its k components by Δ. The first trajectory point, $\mathbf{x}^{(1)}$, is obtained by increasing one or more components of $\mathbf{x}^*$ by Δ, in such a way that $\mathbf{x}^{(1)}$ is still in Ω. The second trajectory point, $\mathbf{x}^{(2)}$, is generated from $\mathbf{x}^*$ with the requirement that it differs from $\mathbf{x}^{(1)}$ in its ith component, which has been either increased or decreased by Δ, i.e. $\mathbf{x}^{(2)} = \mathbf{x}^{(1)} + \mathbf{e}_i \Delta$ or $\mathbf{x}^{(2)} = \mathbf{x}^{(1)} - \mathbf{e}_i \Delta$. The index i is randomly selected in the set $\{1, 2, \ldots, k\}$. The third sampling point, $\mathbf{x}^{(3)}$, is generated from $\mathbf{x}^*$ with the property that $\mathbf{x}^{(3)}$ differs from $\mathbf{x}^{(2)}$ for only one component j, for any $j \neq i$. It can be either $\mathbf{x}^{(3)} = \mathbf{x}^{(2)} + \mathbf{e}_j \Delta$ or $\mathbf{x}^{(3)} = \mathbf{x}^{(2)} - \mathbf{e}_j \Delta$. And so on until $\mathbf{x}^{(k+1)}$, which closes the trajectory. The design produces a trajectory of $(k+1)$ sampling points $\mathbf{x}^{(1)}, \mathbf{x}^{(2)}, \ldots, \mathbf{x}^{(k+1)}$ with the key properties that two consecutive points differ in only one component and that any value of the *base vector* $\mathbf{x}^*$ has been selected at least once to be increased by Δ. An example of a trajectory for $k = 3$ is illustrated in Figure 3.2.

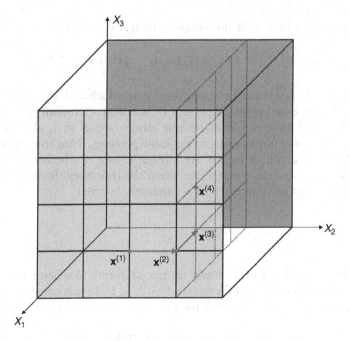

Figure 3.2 An example of a trajectory in the input space when $k = 3$

A technical scheme to generate trajectories with the required properties is as follows. A trajectory can be seen in the form of a matrix, $\mathbf{B}^*$, with dimension $(k+1) \times k$, whose rows are the vectors $\mathbf{x}^{(1)}, \mathbf{x}^{(2)}, \ldots, \mathbf{x}^{(k+1)}$. To build $\mathbf{B}^*$, the first step is the selection of a matrix $\mathbf{B}$, whose dimensions are $(k+1) \times k$, with elements that are 0's and 1's and the key property that for every column index $j, j = 1, \ldots, k$, there are two rows of $\mathbf{B}$ that differ only in the jth entry. A convenient choice for $\mathbf{B}$ is a strictly lower triangular matrix of 1's:

$$\mathbf{B} = \begin{bmatrix} 0 & 0 & 0 & \cdots & 0 \\ 1 & 0 & 0 & \cdots & 0 \\ 1 & 1 & 0 & \cdots & 0 \\ 1 & 1 & 1 & 0 & \cdots \\ \cdots & \cdots & \cdots & \cdots & \cdots \end{bmatrix}.$$

The matrix $\mathbf{B}'$, given by

$$\mathbf{B}' = \mathbf{J}_{k+1,k}\mathbf{x}^* + \Delta \mathbf{B},$$

where $\mathbf{J}_{k+1,k}$ is a $(k+1) \times k$ matrix of 1's and $\mathbf{x}^*$ is a randomly chosen *base value* of $\mathbf{X}$, is a potential candidate for the desired design matrix, but it has the limitation that the kth elementary effect it produces would not be randomly selected.

A randomized version of the sampling matrix is given by

$$\mathbf{B}^* = \left(\mathbf{J}_{k+1,1}\mathbf{x}^* + (\Delta/2) \left[(2\mathbf{B} - \mathbf{J}_{k+1,k}) \mathbf{D}^* + \mathbf{J}_{k+1,k} \right] \right) \mathbf{P}^*, \qquad (3.2)$$

where $\mathbf{D}^*$ is a k-dimensional diagonal matrix in which each element is either $+1$ or -1 with equal probability, and $\mathbf{P}^*$ is a k-by-k random permutation matrix in which each row contains one element equal to 1, all others are 0, and no two columns have 1's in the same position. Read row by row, $\mathbf{P}^*$ gives the order in which factors are moved; $\mathbf{D}^*$ states whether the factors will increase or decrease their value along the trajectory. $\mathbf{B}^*$ provides one elementary effect per input, which is randomly selected.

Example 3.1

Consider a model with two input factors uniformly distributed in $[0, 1]$. Consider the following set of levels $\{0, 1/3, 2/3, 1\}$. In this case $k = 2$, $p = 4$, and we choose $\Delta = 2/3$. Suppose the randomly generated $\mathbf{x}^*$, $\mathbf{D}^*$ and $\mathbf{P}^*$ are

$$\mathbf{x}^* = \{1/3, 1/3\} \qquad \mathbf{D}^* = \begin{bmatrix} 1 & 0 \\ 0 & -1 \end{bmatrix} \qquad \mathbf{P}^* = \mathbf{I}.$$

The matrix **B** is given by

$$\mathbf{B} = \begin{bmatrix} 0 & 0 \\ 1 & 0 \\ 1 & 1 \end{bmatrix}$$

and for these values we get

$$(\Delta/2)\left[(2\mathbf{B} - \mathbf{J}_{k+1,k})\mathbf{D}^* + \mathbf{J}_{k+1,k}\right] = \begin{bmatrix} 0 & \Delta \\ \Delta & \Delta \\ \Delta & 0 \end{bmatrix} = \begin{bmatrix} 0 & 2/3 \\ 2/3 & 2/3 \\ 2/3 & 0 \end{bmatrix}$$

and

$$\mathbf{B}^* = \begin{bmatrix} 1/3 & 1 \\ 1 & 1 \\ 1 & 1/3 \end{bmatrix}$$

so that $\mathbf{x}^{(1)} = (1/3, 1)$; $\mathbf{x}^{(2)} = (1, 1)$; $\mathbf{x}^{(1)} = (1, 1/3)$ (see Figure 3.3). This procedure is repeated a number of times in order to build the distributions F_i and G_i for each factor.

Campolongo *et al.* (2007) proposed an improvement of the sampling strategy just described that facilitates a better scanning of the input domain without increasing the number of model executions needed. The idea is to select the *r* trajectories in such a way as to maximize their spread in the

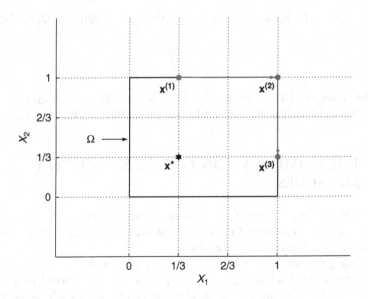

Figure 3.3 The trajectory obtained in Example 3.1

input space. The design starts by generating a high number of different trajectories, e.g. $M \sim 500$–1000, and then selects the subset of r (e.g. $r = 10, 20$) with the highest spread, where the concept of *spread* is based on the following definition of distance, d_{ml}, between a pair of trajectories m and l:

$$d_{ml} = \begin{cases} \sum_{i=1}^{k+1} \sum_{j=1}^{k+1} \sqrt{\sum_{z=1}^{k} \left[X_z^{(i)}(m) - X_z^{(j)}(l) \right]^2} & m \neq l \\ 0 & \text{otherwise} \end{cases} \quad (3.3)$$

where k is the number of input factors and $X_z^{(i)}(m)$ indicates the zth coordinate of the ith point of the mth trajectory. In other words, d_{ml} is the sum of the geometric distances between all the pairs of points of the two trajectories under analysis. The best r trajectories from M are selected by maximizing the distance d_{ml} among them. First we consider for each possible combination of r trajectories from M the quantity D^2, which is the sum of the squared distances between all possible pairs of trajectories belonging to the combination. For instance, for the combination $4, 6, 7$, and 9 (i.e. $r = 4$) from the possible $M = \{1, 2, 3, 4, 5, 6, 7, 8, 9, 10\}$, we define $D_{4,6,7,9}$ as

$$D_{4,6,7,9} = \sqrt{d_{4,6}^2 + d_{4,7}^2 + d_{4,9}^2 + d_{6,7}^2 + d_{6,9}^2 + d_{7,9}^2}.$$

Then, we consider the combination with the highest value of D, of the

$$\binom{10}{4}$$

possible choices of $r = 4$ from $M = 10$. This sampling strategy optimizes the scanning of the input space and is simple to implement.

3.4 THE COMPUTATION OF THE SENSITIVITY MEASURES

HOW CAN I DEFINE A SENSITIVITY MEASURE USING THE ELEMENTARY EFFECTS? HOW MANY LEVELS SHOULD I CHOOSE? HOW CAN I DEAL WITH NON-UNIFORM DISTRIBUTIONS? WHAT IS THE ROLE OF Δ?

The sampling strategy described above results in the construction of r trajectories in Ω. Each trajectory corresponds to $(k+1)$ model executions and allows the computation of an elementary effect for each factor i, for $i = 1, .., k$. If $\mathbf{x}^{(l)}$ and $\mathbf{x}^{(l+1)}$, with l in the set $\{1, \ldots, k\}$, are two sampling points of the *j*th trajectory differing in their *i*th component, the elementary effect associated with factor *i* is

$$EE_i^j\left(\mathbf{x}^{(l)}\right) = \frac{\left[y\left(\mathbf{x}^{(l+1)}\right) - y\left(\mathbf{x}^{(l)}\right)\right]}{\Delta}$$

if the *i*th component of $\mathbf{x}^{(l)}$ is increased by Δ, and

$$EE_i^j\left(\mathbf{x}^{(l+1)}\right) = \frac{\left[y\left(\mathbf{x}^{(l)}\right) - y\left(\mathbf{x}^{(l+1)}\right)\right]}{\Delta}$$

if the *i*th component of $\mathbf{x}^{(l)}$ is decreased by Δ. Once *r* elementary effects per input are available ($EE_i^j, i = 1, 2, \ldots, k, j = 1, 2, \ldots, r$), the statistics μ_i, μ_i^* and σ_i^2 relative to the distributions F_i and G_i can be computed for each factor by using the same estimators that would be used with independent random samples, as the *r* elementary effects belong to different trajectories and are therefore independent:

$$\mu_i = \frac{1}{r}\sum_{j=1}^{r} EE_i^j \tag{3.4}$$

$$\mu_i^* = \frac{1}{r}\sum_{j=1}^{r} |EE_i^j| \tag{3.5}$$

$$\sigma_i^2 = \frac{1}{r-1}\sum_{j=1}^{r}\left(EE_i^j - \mu\right)^2 \tag{3.6}$$

where EE_i^j indicates the elementary effects relative to factor *i* computed along trajectory *j*.

A critical choice related to the implementation of the method is the choice of the parameters p and Δ. The choice of p is strictly linked to the choice of r. If one considers a high value of p, thus producing a high number of possible levels to be explored, one is only seemingly enhancing the accuracy of the sampling. If this is not coupled with the choice of a high value of r, the effort will be wasted, since many possible levels will remain unexplored. In general, when the sampling size r is small, it is likely that not all possible factor levels will be explored within the experiment. For instance, in the above example, if $r = 1$ the two factors never take the values 0 and 2/3. If possible, it is convenient to choose an even value for p as this affects the choice of Δ. In fact, assuming p to be even, a convenient choice for Δ is $\Delta = p/[2(p-1)]$. This choice has the advantage that the

ELEMENTARY EFFECTS METHOD

design's sampling strategy guarantees equal-probability sampling from each F_i (for details see Morris, 1991).

The top part of Figure 3.4 (grey arrows) shows that when $p = 4$, the choice of $\Delta = p/[2(p-1)] = 2/3$ (left plot) guarantees that the four levels have equal probability of being selected. On the other hand, a choice of $\Delta = 1/3$ (right plot) would imply that the levels $1/3$ and $2/3$ are sampled more often, since there are two arrays pointing there. The two histograms

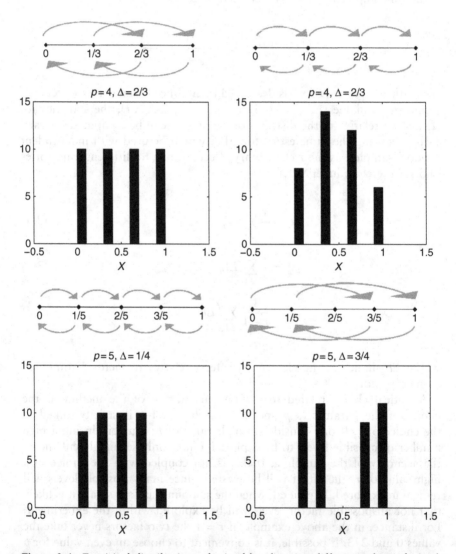

Figure 3.4 Empirical distributions obtained by choosing different values of p and Δ and sampling $r = 20$ trajectories

THE COMPUTATION OF THE SENSITIVITY MEASURES

below the arrows plot the empirical distributions obtained when generating $r = 20$ trajectories for $p = 4$ levels, in the case of $\Delta = 2/3$ (left plot) and $\Delta = 1/3$ (right plot). The bottom part of the figure illustrates the case in which an odd number of levels is considered ($p = 5$). Under this condition, no matter what value of Δ is chosen, it is impossible to achieve equal probability for the elementary effects. In some cases, e.g. for $p = 5$ and $\Delta = 3/4$, there are elementary effects which can never be sampled. Previous experiments (Campolongo and Saltelli, 1997; Campolongo et al., 1999b; Saltelli et al., 2000) have demonstrated that the choice of $p = 4$ and $r = 10$ has produced valuable results.

If a factor follows a uniform distribution, the levels are obtained simply by dividing into equal parts the interval in which each factor varies. If a factor follows nonuniform distributions, then input values are not sampled directly; rather, the sampling is carried out in the space of the quantiles of the distributions, which is a k-dimensional hypercube (each quantile varies in [0, 1]). The actual factor values are then derived from its known statistical distribution (Campolongo et al., 1999b). Figure 3.5 shows the case where a factor X follows a normal standard distribution and the space of quantiles is investigated via six quantiles ($q1, q2, \ldots, q6$). Using the inverse of the normal cumulative distribution function (grey curve), corresponding levels for X are obtained ($L1, L2, \ldots, L6$). If the distribution to be sampled has an infinite support, the quantiles to be used can be chosen for instance cutting the tails of the distribution. A good alternative is to divide the entire

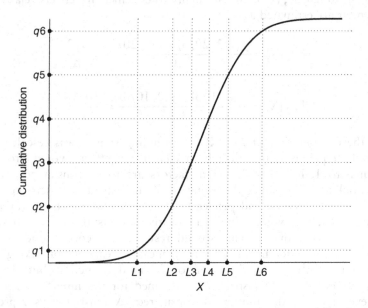

Figure 3.5 Sampling procedure for factors with standard normal distribution

support into a number of intervals equal to $(r+1)$ and then to use the centres of the bins (see also Section 2.3.2).

At this point, it is worth considering briefly the role of Δ. If Δ is chosen to be equal for all input factors, its role in the definition of the elementary effect for sensitivity purposes becomes irrelevant. Δ^{-1} is simply a constant multiplying each elementary effect, which does not affect the sensitivity analysis results. A question may arise in the case that an input factor needs to be rescaled because its original distribution is not uniform between 0 and 1: *Should Δ be rescaled so as to represent the actual sampling step in the factor range of variation, or should it remain the same so as to represent the sampling step between [0, 1] equal for each factor?* The following example addresses this question.

Assume the output Y is the simple function $Y = X_1 + X_2$, with X_1 and X_2 uniformly distributed in $[0, 1]$ and $[0, 10]$ respectively. In our sensitivity results one expects that input factor X_2 is much more important than input factor X_1, since a variation in its value affects Y much more than a variation in X_1, i.e. X_2 is much more uncertain than X_1. Here we consider $p = 4$ levels, $\Delta = 1/3$, and $r = 1$ trajectory, i.e. we compute just one elementary effect for each input and consider these as the final sensitivity measures. Assume that we randomly generate the following trajectory of quantiles for X_1 and X_2: $(0, 1/3)$; $(0, 2/3)$; $(1/3, 2/3)$. The reader familiar with the method can easily verify that this is a typical sample. Applying the inverse cumulative function we obtain this sample for the two factors: $(0, 10/3)$; $(0, 20/3)$; $(1/3, 20/3)$. Finally, the elementary effects relative to each input are estimated as

$$EE_1(\mathbf{X}) = \frac{y(1/3, 20/3) - y(0, 20/3)}{\Delta} = \frac{1/3}{\Delta}$$

$$EE_2(\mathbf{X}) = \frac{y(0, 20/3) - y(0, 10/3)}{\Delta} = \frac{10/3}{\Delta}.$$

If in the computation of $EE_2(\mathbf{X})$ the sampling step Δ was rescaled and set to be equal to $10/3$, then both elementary effects would prove to be equal to 1, implying that both factors are to be considered equally influential in determining the value of Y. If instead Δ is left equal to $1/3$ for both factors, independently of the actual range of variation of factor X_2, we would obtain a sensitivity measure for X_2 10 times higher than that of X_1. This second result is the correct one. In this way we are considering a sensitivity measure capable of taking into account not only the specifications of the model function, but also the effect of the statistical distributions assumed for the inputs. In general, whatever type of distribution is considered, Δ must always represent the sampling step in the scale $[0, 1]$, i.e. in the case of non-uniform distributions it should represent the variation in the quantiles of the factors.

3.5 WORKING WITH GROUPS

HOW CAN I EXTEND THE EE METHOD TO DEAL WITH GROUPS OF FACTORS? DOES THE SAMPLING STRATEGY CHANGE?

The EE method presented above can also be extended to work with groups using the μ^* measure alone. When working with groups, the idea is to move all factors of the same group simultaneously. In the original definition given by Morris, the elementary effect is obtained by subtracting the function evaluated at X from that evaluated after incrementing one factor see Equation (3.1). This definition cannot be extended straightforwardly to cases in which more than one factor is moved at the same time, as two factors may have been changed in opposite directions, i.e. one increased and one decreased by Δ. By contrast, using μ^* overcomes this problem, as the focus is not on the elementary effect itself but on its absolute value, i.e. the elementary effect is always positive, regardless of the displacement of the factors. For a two-factor group $\mathbf{u} = (X_{i_1}, X_{i_2})$, the absolute elementary effect in point X is

$$|EE_\mathbf{u}(\mathbf{X})| = \left| \frac{y(\tilde{\mathbf{X}}) - y(\mathbf{X})}{\Delta} \right| \qquad (3.7)$$

where $\mathbf{X}$ is any selected value in Ω such that the transformed point $\tilde{\mathbf{X}}$ is still in Ω, and each of the components $(\tilde{X}_{i_1}, \tilde{X}_{i_2})$ has been either increased or decreased by Δ with respect to (X_{i_1}, X_{i_2}).

In order to treat groups of factors, the sampling strategy described in Section 3.3 needs to be modified slightly. It is first necessary to consider a matrix $\mathbf{G}$ describing how factors are allocated into groups. This matrix is defined as follows: its element $G(i,j)$ equals 1 if factor i belongs to group j; otherwise $G(i,j) = 0$. If g is the number of groups in the experiment, $\mathbf{G}$ has sizes $k \times g$. In this case the matrix of trajectories $\mathbf{B}^*$ has dimensions $(g+1) \times k$, since all the factors in a group move together. $\mathbf{B}^*$ can be built considering a lower triangular matrix $\mathbf{B}$ whose dimensions are $(g+1) \times g$ and setting

$$\mathbf{B}^* = \mathbf{J}_{g+1,1}\mathbf{x}^* + (\Delta/2)\left[\left(2\mathbf{B}(\mathbf{G}\mathbf{P}^*)^T - \mathbf{J}_{g+1,k}\right)\mathbf{D}^* + \mathbf{J}_{g+1,k}\right],$$

where, similarly to the single-factor experiment, $\mathbf{J}_{i,j}$ is a matrix of 1's with dimensions $(i \times j)$; $\mathbf{D}^*$ is a diagonal matrix $(k \times k)$ describing whether the factors increase or decrease value; and $\mathbf{P}^*$ is a $(g \times g)$ matrix, describing the order in which the groups move. The following example illustrates how to handle groups of factors.

Example 3.2

Consider three factors X_1, X_2 and X_3, uniformly distributed on $[0, 1]$, and assigned to a number of groups $g = 2$. The first group (G_1) contains only factor X_1; the second (G_2) contains the other two factors. The matrix **G** is therefore defined as

$$\mathbf{G} = \begin{bmatrix} 1 & 0 \\ 0 & 1 \\ 0 & 1 \end{bmatrix}.$$

Consider an experiment with $p = 4$ levels and the choice $\Delta = 2/3$. Suppose we obtain the following matrices for $\mathbf{x}^*$, $\mathbf{D}^*$ and $\mathbf{P}^*$:

$$\mathbf{x}^* = (1/3, 1/3, 0) \qquad \mathbf{D}^* = \begin{bmatrix} 1 & 0 & 0 \\ 0 & -1 & 0 \\ 0 & 0 & 1 \end{bmatrix} \qquad \mathbf{P}^* = \begin{bmatrix} 1 & 0 \\ 0 & 1 \end{bmatrix}.$$

In this case we get

$$(\Delta/2)\left[\left(2\mathbf{B}(\mathbf{GP}^*)^T - \mathbf{J}_{g+1,k}\right)\mathbf{D}^* + \mathbf{J}_{g+1,k}\right]$$

$$= \Delta/2 \left[\left(2 \begin{bmatrix} 0 & 0 \\ 1 & 0 \\ 1 & 1 \end{bmatrix} \begin{bmatrix} 1 & 0 \\ 0 & 1 \\ 0 & 1 \end{bmatrix} - \begin{bmatrix} 1 & 1 & 1 \\ 1 & 1 & 1 \\ 1 & 1 & 1 \end{bmatrix} \right) \begin{bmatrix} 1 & 0 & 0 \\ 0 & -1 & 0 \\ 0 & 0 & 1 \end{bmatrix} + \begin{bmatrix} 1 & 1 & 1 \\ 1 & 1 & 1 \\ 1 & 1 & 1 \end{bmatrix} \right]$$

$$= \begin{bmatrix} 0 & \Delta & 0 \\ \Delta & \Delta & 0 \\ \Delta & 0 & \Delta \end{bmatrix}$$

which makes clear that X_1 (i.e. G_1) moves first and increases in value, and then that the factors in G_2 change their values in opposite directions (X_2 decreases and X_3 increases). The final matrix is then:

$$\mathbf{B}^* = \begin{bmatrix} 1/3 & 1 & 0 \\ 1 & 1 & 0 \\ 1 & 1/3 & 2/3 \end{bmatrix}.$$

Note that the two factors in the same group also take values on different levels.

3.6 THE EE METHOD STEP BY STEP

In this section we show how to put the EE method illustrated above into practice. The method is tested on the analytical g-function attributed to Sobol' (1990):

$$Y = \prod_{i=1}^{k} g_i(X_i) \quad \text{where} \quad g_i(X_i) = \frac{|4X_i - 2| + a_i}{1 + a_i}$$

and a_i are parameters, such that $a_i \geq 0$. This function is widely used as a test function in sensitivity analysis because it is a very difficult one: it is strongly nonlinear and nonmonotonic, and all its interaction terms are nonzero by definition. Moreover it is possible to compute analytically the output variance decomposition and therefore the variance-based sensitivity indices S_i and S_{T_i}. The values of the a_i determine the relative importance of the X_i as they determine the range of variation of each $g_i(X_i)$:

$$1 - \frac{1}{1+a_i} \leq g_i(X_i) \leq 1 + \frac{1}{1+a_i}.$$

Thus, the higher the a_i value, the lower the importance of the X_i variable. This is also shown by Figure 3.6, which illustrates the behaviour of $g_i(X_i)$ as a function of X_i for the values of $a_i = 0.9$, $a_i = 9$ and $a_i = 99$.

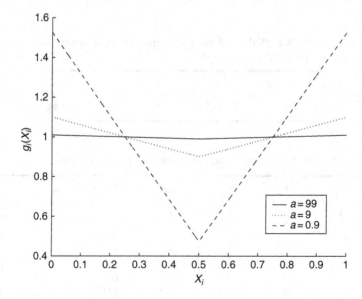

Figure 3.6 $g_i(X_i)$ as a function of X_i for $a_i = 0.9$, $a_i = 9$ and $a_i = 99$

The partial variances of the first order are given by

$$V_i = V(E(Y|X_i)) = \frac{1}{3(1+a_i)^2},$$

while the higher-order partial variances are simply the product of the lower ones, i.e. $V_{12} = V_1 V_2$. Finally, the total variance can be written as a function of the first-order terms only:

$$V(Y) = -1 + \prod_{i=1}^{k}(1+V_i).$$

This allows us to calculate analytically the first- and total-order sensitivity indices for the g-function. In our example we assume $k = 6$ and we set the a_i parameter values as shown in Table 3.1.

We now perform the sensitivity analysis by making use of the EE method introduced in this chapter. The uncertainty in the input variables is modelled by assuming that all the X_i are uniformly distributed in the six-dimensional unit cube (i.e. $X_i \sim U[0, 1], \forall i$). The optimized strategy described in Section 3.3 is used to estimate the sensitivity measures μ, μ^* and σ. $p = 4$ levels and $\Delta = 2/3$ are chosen; $r = 4$ trajectories are employed, selected out of a group of 100. The sampled input matrix is presented in Table 3.2, columns 2–7. The last column reports the corresponding values for the g-function. The graphs in Figure 3.7 present the scatterplots of the

Table 3.1 Values of the parameters of the g-function for each input

a_1	a_2	a_3	a_4	a_5	a_6
78	12	0.5	2	97	33

Table 3.2 Sampled trajectories and corresponding g-function values

	X_1	X_2	X_3	X_4	X_5	X_6	g
	0	2/3	1	0	0	1/3	2.193
	0	2/3	1	0	0	1	2.280
	0	0	1	0	0	1	2.520
t_1	2/3	0	1	0	0	1	2.478
	2/3	0	1	2/3	0	1	1.652
	2/3	0	1/3	2/3	0	1	0.771
	2/3	0	1/3	2/3	2/3	1	0.761

	0	1/3	1/3	1	1	2/3	1.024
	0	1	1/3	1	1	2/3	1.131
	0	1	1	1	1	2/3	2.424
t_2	2/3	1	1	1	1	2/3	2.384
	2/3	1	1	1	1	0	2.478
	2/3	1	1	1	1/3	0	2.445
	2/3	1	1	1/3	1/3	0	1.630
	1	2/3	0	2/3	1	0	1.520
	1	2/3	0	0	1	0	2.280
	1/3	2/3	0	0	1	0	2.242
t_3	1/3	2/3	0	0	1/3	0	2.212
	1/3	0	0	0	1/3	0	2.445
	1/3	0	2/3	0	1/3	0	1.141
	1/3	0	2/3	0	1/3	2/3	1.097
	1	1/3	2/3	1	0	1/3	1.024
	1	1/3	2/3	1	0	1	1.064
	1	1/3	0	1	0	1	2.280
t_4	1	1/3	0	1/3	0	1	1.520
	1	1/3	0	1/3	2/3	1	1.500
	1	1	0	1/3	2/3	1	1.657
	1/3	1	0	1/3	2/3	1	1.630

output as a function of each input. In this example, in which the number of inputs is not too large, the use of scatterplots already reveals that one factor (X_3) plays a key role in the function. For more complex examples, in which the number of factors is larger, the use of scatterplots is cumbersome.

As an example, Table 3.3 shows how to estimate the sensitivity measures for factor X_4. The values of μ, μ^* and σ are reported in Table 3.4. Results indicate that X_3 and X_4 are important factors, while factors X_1, X_5 and X_6 can be regarded as noninfluential (see the μ^* values in the first column). The high values of σ for some factors also demonstrate that interactions play an important role in the model. Moreover, the low values of μ associated with high values of μ^* indicate that factors have effects of oscillating signs. Figure 3.8 shows in a barplot the values of μ^* in increasing order. It is clear that in this example, μ^* alone can be used to assess the importance of each factor of the model. These results can be used to fix unimportant factors in the model (the *Factor Fixing* setting, see also Chapters 1 and 4). The extremely low values of μ^* for X_1 and X_5 show that these factors can be fixed without affecting the variance of g to a great extent. On the other hand, it is clear that factors X_3 and X_4 cannot be fixed. The remaining factors (X_2 and X_6) stay in-between. The decision to fix them or not depends on how much importance is assigned to type II errors.

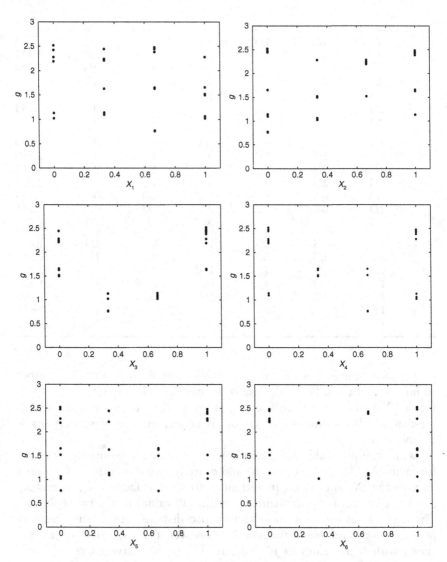

Figure 3.7 Scatterplot of the $g = \prod_{i+1}^{k} g_i(X_i)$ as a function of each input

Table 3.3 Estimation of the distribution of the elementary effects (and their absolute values) for factor X_4

	$g(X_1,\ldots,X_4+\Delta,\ldots,X_6)$	$g(X_1,\ldots,X_4,\ldots,X_6)$	$EE_4(X)$	$|EE_4(X)|$
t_1	1.652	2.478	−1.239	1.239
t_2	2.445	1.630	1.222	1.222
t_3	1.520	2.280	−1.140	1.140
t_4	2.280	1.520	1.140	1.140

CONCLUSIONS

Table 3.4 Estimated sensitivity measures. The measures are estimated using $r = 4$ trajectories

	μ^*	μ	σ
X_1	0.056	−0.006	0.064
X_2	0.277	−0.078	0.321
X_3	1.760	−0.130	2.049
X_4	1.185	−0.004	1.370
X_5	0.035	0.012	0.041
X_6	0.099	−0.004	0.122

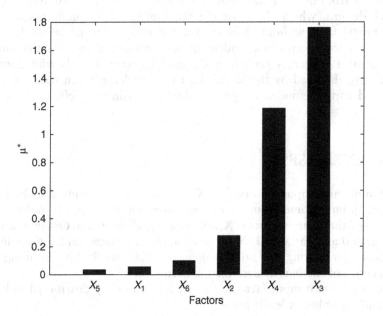

Figure 3.8 Barplot of μ^* for the g-function

3.7 CONCLUSIONS

The EE method illustrated in this chapter is effective in identifying the few important factors in a model that contains many factors, with a relatively small number of model evaluations. The method is convenient when the number of factors is large and/or the model execution time is such that the computational cost of more sophisticated techniques is excessive, but not large enough to oblige the modeller to make use of grouping techniques.

With respect to a group technique, the EE method has the advantage of examining each factor individually so as to prevent cancellation effects and to offer a well-defined strategy to assess the importance of each individual factor. The method is conceptually simple. It can be thought of as an expansion of a derivative-based approach: when a single trajectory is considered, and the variations of input factors are small, it reduces to a incremental ratio estimation. Nevertheless, it overcomes the limitations of a local derivative-based approach in that it attempts to explore the whole input space. The method is computationally easy and has the flexibility typical of OAT approaches in handling unstable models. When a model is unstable it threatens to crash if executed on a set of input values rather different from the nominal values on which it was calibrated. Similarly to OAT designs, which allow for the substitution of a sample point when the model fails, without changing the full design, the EE method, being based on trajectories independent of one another, allows for the substitution of a trajectory on which the model execution fails with another one better handled by the model. Last but not least, when necessary, the method can be applied to groups, thus increasing the efficiency of the design.

3.8 EXERCISES

1. Given three input factors, X_1, X_2 and X_3, all uniformly distributed in $[0, 1]$, build some examples of trajectories for $p = 4$, $p = 6$, and $p = 8$.
2. Given three input factors, X_1, X_2 and X_3, all normally distributed with mean 0 and variance 1, build an example of trajectories for $p = 6$ levels (consider cutting the tails at quantiles 0.5 and 99.5%, or using the centres of the bins as described in Section 2.4.3).
3. Consider two input factors, X_1 and X_2, uniformly distributed in $[0, 1]$ and a number of levels $p = 4$.

    (a) Following the revised sampling strategy, generate randomly a number $M = 6$ of trajectories and then select, according to the approach described in Section 3.3, the best four out of six.
    (b) Show by means of graphical representation that the new strategy outperforms the old one in terms of input-space scanning

4. As a follow-up to Exercise 3, given the model

$$y = \sin(\pi X_1) + \cos(\pi X_2/4) + \sqrt{X_1 X_2},$$

make use of the full set of $M = 6$ trajectories and compute the sensitivity measures μ, μ^* and σ for the two input factors X_1 and X_2.

EXERCISES

5. Consider a model with 15 input factors with standard normal distribution, i.e. $X_i \sim N(0, 1)$. Consider the function (Oakley and O'Hagan, 2004)

$$y = \mathbf{a}_1^T + \mathbf{a}_2^T \cos(\mathbf{X}) + \mathbf{a}_3^T \sin(\mathbf{X}) + \mathbf{X}^T \mathbf{M} \mathbf{X}$$

where $\mathbf{a}_j$, ($j = 1, 2, 3$) and $\mathbf{M}$ are respectively three vectors and a matrix of parameters, whose values are reported in Tables 3.5 and 3.6. Consider cutting the tails of the normal distributions at quantiles 5 and 95%. For this model estimate the sensitivity measures μ^*, μ.

6. As a follow-up to Exercise 5, group the 15 factors into the two groups described by the matrix $\mathbf{G}$ below and compute the sensitivity measures for the groups using $r = 10$ trajectories.

$$\mathbf{G}^T = \begin{bmatrix} 0 & 0 & 0 & 0 & 0 & 0 & 0 & 1 & 0 & 1 & 1 & 1 & 1 & 1 & 1 \\ 1 & 1 & 1 & 1 & 1 & 1 & 1 & 0 & 1 & 0 & 0 & 0 & 0 & 0 & 0 \end{bmatrix}$$

7. Consider the function introduced by Morris (1991):

$$y = \beta_0 + \sum_{i=1}^{20} \beta_i \omega_i + \sum_{i<j}^{20} \beta_{i,j} \omega_i \omega_j + \sum_{i<j<l}^{20} \beta_{i,j,l} \omega_i \omega_j \omega_l + \sum_{i<j<l<s}^{20} \beta_{i,j,l,s} \omega_i \omega_j \omega_l \omega_s$$

where $\omega_i = 2(X_i - 1/2)$ except for $i = 3, 5$, and 7, where

$$\omega_i = 2(1.1 X_i/(X_i + 0.1) - 1/2).$$

Table 3.5 Vectors of parameters for the Oakley–O'Hagan function

Factor	a_1	a_2	a_3
X_1	0.01	0.43	0.10
X_2	0.05	0.09	0.21
X_3	0.23	0.05	0.08
X_4	0.04	0.32	0.27
X_5	0.12	0.15	0.13
X_6	0.39	1.04	0.75
X_7	0.39	0.99	0.86
X_8	0.61	0.97	1.03
X_9	0.62	0.90	0.84
X_{10}	0.40	0.81	0.80
X_{11}	1.07	1.84	2.21
X_{12}	1.15	2.47	2.04
X_{13}	0.79	2.39	2.40
X_{14}	1.12	2.00	2.05
X_{15}	1.20	2.26	1.98

Table 3.6 Matrix of parameters for the Oakley–O'Hagan function

	X_1	X_2	X_3	X_4	X_5	X_6	X_7	X_8	X_9	X_{10}	X_{11}	X_{12}	X_{13}	X_{14}	X_{15}
X_1	−0.02	−0.19	0.13	0.37	0.17	0.14	−0.44	−0.08	0.71	−0.44	0.5	−0.02	−0.05	0.22	0.06
X_2	0.26	0.05	0.26	0.24	−0.59	−0.08	−0.29	0.42	0.5	0.08	−0.11	0.03	−0.14	−0.03	−0.22
X_3	−0.06	0.2	0.1	−0.29	−0.14	0.22	0.15	0.29	0.23	−0.32	−0.29	−0.21	0.43	0.02	0.04
X_4	0.66	0.43	0.3	−0.16	−0.31	−0.39	0.18	0.06	0.17	0.13	−0.35	0.25	−0.02	0.36	−0.33
X_5	−0.12	0.12	0.11	0.05	−0.22	0.19	−0.07	0.02	−0.1	0.19	0.33	0.31	−0.08	−0.25	0.37
X_6	−0.28	−0.33	−0.1	−0.22	−0.14	−0.14	−0.12	0.22	−0.03	−0.52	0.02	0.04	0.36	0.31	0.05
X_7	−0.08	0.004	0.89	−0.27	−0.08	−0.04	−0.19	−0.36	−0.17	0.09	0.4	−0.06	0.14	0.21	−0.01
X_8	−0.09	0.59	0.03	−0.03	−0.24	−0.1	0.03	0.1	−0.34	0.01	−0.61	0.08	0.89	0.14	0.15
X_9	−0.13	0.53	0.13	0.05	0.58	0.37	0.11	−0.29	−0.57	0.46	−0.09	0.14	−0.39	−0.45	−0.15
X_{10}	0.06	−0.32	0.09	0.07	−0.57	0.53	0.24	−0.01	0.07	0.08	−0.13	0.23	0.14	−0.45	−0.56
X_{11}	0.66	0.35	0.14	0.52	−0.28	−0.16	−0.07	−0.2	0.07	0.23	−0.04	−0.16	0.22	0	−0.09
X_{12}	0.32	−0.03	0.13	0.13	0.05	−0.17	0.18	0.06	−0.18	−0.31	−0.25	0.03	−0.43	−0.62	−0.03
X_{13}	−0.29	0.03	0.03	−0.12	0.03	−0.34	−0.41	0.05	−0.27	−0.03	0.41	0.27	0.16	−0.19	0.02
X_{14}	−0.24	−0.44	0.01	0.25	0.07	0.25	0.17	0.01	0.25	−0.15	−0.08	0.37	−0.3	0.11	−0.76
X_{15}	0.04	−0.26	0.46	−0.36	−0.95	−0.17	0.003	0.05	0.23	0.38	0.46	−0.19	0.01	0.17	0.16

Each input factor X_i is uniformly distributed in $[0, 1]$. Coefficients with relatively large values are: $\beta_i = 20$, $i = 1, .., 10$; $\beta_{i,j} = -15$, $i, j = 1, .., 6$; $\beta_{i,j,l} = -10$, $i, j, l = 1, .., 5$; $\beta_{i,j,l,s} = 5$, $i, j, l, s = 1, .., 4$. The remaining first- and second-order coefficients are independently generated from a normal distribution with zero mean and unit standard deviation. The remaining third- and fourth-order coefficients are set to zero. For this test case, estimate the sensitivity analysis measures μ^*, μ and σ using the optimized sampling strategy with $r = 10$ trajectories and discuss the results, considering also the structure of the function y.

8. As a follow-up to Exercise 7, group the 20 factors into $g = 4$ groups described by the matrix $\mathbf{G}$ below and compute the sensitivity measures for the groups using $r = 10$ trajectories. Discuss the results, using the sensitivity information from Exercise 7.

$$\mathbf{G}^T = \begin{bmatrix} 1 1 1 0 0 0 0 0 0 1 1 1 0 0 0 0 0 0 0 0 \\ 0 0 0 1 1 0 0 0 0 0 0 0 0 0 0 0 0 0 1 1 \\ 0 0 0 0 0 0 0 0 0 0 0 0 0 1 1 1 1 0 0 0 \\ 0 0 0 0 0 1 1 1 1 0 0 0 1 0 0 0 0 1 0 0 \end{bmatrix}.$$

3.9 SOLUTIONS

1. We treat separately the three cases $p = 4, 6, 8$.

   - **Number of levels $p = 4$**
     The levels are $\{0, 1/3, 2/3, 1\}$. Consider the optimal value for Δ, which is $\Delta = 2/3$, and suppose that the randomly generated $\mathbf{x}^*$, $\mathbf{D}^*$ and $\mathbf{P}^*$ are

     $$\mathbf{x}^* = [0, 1/3, 1/3] \quad \mathbf{D}^* = \begin{bmatrix} -1 & 0 & 0 \\ 0 & 1 & 0 \\ 0 & 0 & 1 \end{bmatrix} \quad \mathbf{P}^* = \begin{bmatrix} 0 & 0 & 1 \\ 1 & 0 & 0 \\ 0 & 1 & 0 \end{bmatrix}.$$

     For these values,

     $$(\Delta/2)\left[(2\mathbf{B} - \mathbf{J}_{k+1,k})\mathbf{D}^* + \mathbf{J}_{k+1,k}\right] = \begin{bmatrix} \Delta & 0 & 0 \\ 0 & 0 & 0 \\ 0 & \Delta & 0 \\ 0 & \Delta & \Delta \end{bmatrix}$$

and then
$$B^* = \begin{bmatrix} 1/3 & 1/3 & 2/3 \\ 1/3 & 1/3 & 0 \\ 1 & 1/3 & 0 \\ 1 & 1 & 0 \end{bmatrix}.$$

The trajectory obtained is $x^{(1)} = (1/3, 1/3, 2/3)$; $x^{(2)} = (1/3, 1/3, 0)$; $x^{(3)} = (1, 1/3, 0)$; $x^{(4)} = (1, 1, 0)$.

- **Number of levels $p = 6$**

  The levels are $\{0, 1/5, 2/5, 3/5, 4/5, 1\}$. Suppose that $\Delta = 3/5$, and suppose that the randomly generated x^*, D^* and P^* are $x^* = [0, 1/5, 2/5]$;

  $$D^* = \begin{bmatrix} -1 & 0 & 0 \\ 0 & -1 & 0 \\ 0 & 0 & -1 \end{bmatrix} \quad P^* = \begin{bmatrix} 0 & 1 & 0 \\ 0 & 0 & 1 \\ 1 & 0 & 0 \end{bmatrix}.$$

  For these values,

  $$(\Delta/2)\left[(2B - J_{k+1,k})D^* + J_{k+1,k}\right] = \begin{bmatrix} \Delta & \Delta & \Delta \\ 0 & \Delta & \Delta \\ 0 & 0 & \Delta \\ 0 & 0 & 0 \end{bmatrix}$$

  and then

  $$B^* = \begin{bmatrix} 1 & 3/5 & 4/5 \\ 1 & 0 & 4/5 \\ 1 & 0 & 1/5 \\ 2/5 & 0 & 1/5 \end{bmatrix}.$$

  The trajectory obtained is $x^{(1)} = (1, 3/5, 4/5)$; $x^{(2)} = (1, 0, 4/5)$; $x^{(3)} = (1, 0, 1/5)$; $x^{(4)} = (2/5, 0, 1/5)$.

- **Number of levels $p = 8$**

  The levels are $\{0, 1/7, 2/7, 3/7, 4/7, 5/7, 6/7, 1\}$. Suppose that $\Delta = 3/7$, and suppose that the randomly generated x^*, D^* and P^* are $x^* = [1/7, 3/7, 2/7]$;

  $$D^* = \begin{bmatrix} 1 & 0 & 0 \\ 0 & 1 & 0 \\ 0 & 0 & -1 \end{bmatrix} \quad P^* = \begin{bmatrix} 1 & 0 & 0 \\ 0 & 0 & 1 \\ 0 & 1 & 0 \end{bmatrix}.$$

  For these values,

  $$(\Delta/2)\left[(2B - J_{k+1,k})D^* + J_{k+1,k}\right] = \begin{bmatrix} 0 & 0 & \Delta \\ \Delta & 0 & \Delta \\ \Delta & \Delta & \Delta \\ \Delta & \Delta & 0 \end{bmatrix}$$

SOLUTIONS

and then
$$B^* = \begin{bmatrix} 1/7 & 5/7 & 3/7 \\ 4/7 & 5/7 & 3/7 \\ 4/7 & 5/7 & 6/7 \\ 4/7 & 2/7 & 6/7 \end{bmatrix}.$$

The trajectory obtained is $x^{(1)} = (1/7, 5/7, 3/7)$; $x^{(2)} = (4/7, 5/7, 3/7)$; $x^{(3)} = (4/7, 5/7, 6/7)$; $x^{(4)} = (4/7, 2/7, 6/7)$.

2. • **Cut off the tails**

   Since the factors are not uniformly distributed, the quantiles are first sampled and the corresponding values for the factors are obtained subsequently. In this exercise we cut 0.5% of the distribution in each tail. We first sample the quantiles:

   $$\{0.0, 0.2, 0.4, 0.6, 0.8, 1.0\}.$$

   Suppose that $\Delta = 0.6$ and that the randomly generated x^*, D^* and P^* are

   $$x^* = [0.2, 0.4, 0.4] \quad D^* = \begin{bmatrix} 1 & 0 & 0 \\ 0 & -1 & 0 \\ 0 & 0 & -1 \end{bmatrix} \quad P^* = \begin{bmatrix} 0 & 0 & 1 \\ 1 & 0 & 0 \\ 0 & 1 & 0 \end{bmatrix}.$$

   For these values,

   $$B^* = \begin{bmatrix} 1.0 & 1.0 & 0.2 \\ 1.0 & 1.0 & 0.8 \\ 0.4 & 1.0 & 0.8 \\ 0.4 & 0.4 & 0.8 \end{bmatrix}.$$

   To cut the tails, we rescale these values in the range $[0.005, 0.995]$, thus obtaining

   $$B^* = \begin{bmatrix} 0.995 & 0.995 & 0.203 \\ 0.995 & 0.995 & 0.797 \\ 0.401 & 0.995 & 0.797 \\ 0.401 & 0.401 & 0.797 \end{bmatrix}.$$

   Finally, by applying the inverse normal cumulative function Φ^{-1}, the final values for the factors are obtained:

   $$\Phi^{-1} B^* = \begin{bmatrix} 2.576 & 2.576 & -0.831 \\ 2.576 & 2.576 & 0.831 \\ -0.251 & 2.576 & 0.831 \\ -0.251 & -0.251 & 0.831 \end{bmatrix}.$$

- **Centres of bins**
  In this case we divide the interval [0, 1] into six subintervals and we consider the centres of the bins as levels:

  $$\{0.083, 0.250, 0.417, 0.583, 0.750, 0.917\}.$$

  Suppose that $\Delta = 0.5$, $\mathbf{x}^* = [0.083, 0.417, 0.250]$, and that the randomly generated $\mathbf{D}^*$ and $\mathbf{P}^*$ are as above. Then we obtain

  $$\mathbf{B}^* = \begin{bmatrix} 0.917 & 0.750 & 0.083 \\ 0.917 & 0.750 & 0.583 \\ 0.417 & 0.750 & 0.583 \\ 0.417 & 0.250 & 0.583 \end{bmatrix}.$$

  By applying the inverse normal cumulative function Φ^{-1}, the corresponding values for the factors are obtained:

  $$\Phi^{-1}\mathbf{B}^* = \begin{bmatrix} 1.383 & 0.675 & -1.383 \\ 1.383 & 0.675 & 0.210 \\ -0.210 & 0.675 & 0.210 \\ -0.210 & -0.675 & 0.210 \end{bmatrix}.$$

3. (a) Consider generating the following six trajectories:

$$\mathbf{B}^*_1 = \begin{bmatrix} 0 & 1/3 \\ 0 & 1 \\ 2/3 & 1 \end{bmatrix} \quad \mathbf{B}^*_2 = \begin{bmatrix} 0 & 1/3 \\ 2/3 & 1/3 \\ 2/3 & 1 \end{bmatrix} \quad \mathbf{B}^*_3 = \begin{bmatrix} 2/3 & 0 \\ 2/3 & 2/3 \\ 0 & 2/3 \end{bmatrix}$$

$$\mathbf{B}^*_4 = \begin{bmatrix} 1/3 & 1 \\ 1 & 1 \\ 1 & 1/3 \end{bmatrix} \quad \mathbf{B}^*_5 = \begin{bmatrix} 1/3 & 1 \\ 1/3 & 1/3 \\ 1 & 1/3 \end{bmatrix} \quad \mathbf{B}^*_6 = \begin{bmatrix} 1/3 & 2/3 \\ 1/3 & 0 \\ 1 & 0 \end{bmatrix}.$$

Following the notation introduced in Section 3.3, we denote with $X_z^{(i)}(m)$ the z-component ($z = 1, 2$) of the ith point ($i = 1, 2, 3$) of the mth trajectory ($m = 1, 2, \ldots, 6$). We show, for instance, how to calculate the distance between trajectories $\mathbf{B}^*_1$ and $\mathbf{B}^*_3$:

$$EE_{13} = \sqrt{\left[X_1^{(1)}(1) - X_1^{(1)}(3)\right]^2 + \left[X_2^{(1)}(1) - X_2^{(1)}(3)\right]^2}$$
$$+ \sqrt{\left[X_1^{(1)}(1) - X_1^{(2)}(3)\right]^2 + \left[X_2^{(1)}(1) - X_2^{(2)}(3)\right]^2}$$
$$+ \sqrt{\left[X_1^{(1)}(1) - X_1^{(3)}(3)\right]^2 + \left[X_2^{(1)}(1) - X_2^{(3)}(3)\right]^2}$$

SOLUTIONS

$$+ \sqrt{\left[X_1^{(2)}(1) - X_1^{(1)}(3)\right]^2 + \left[X_2^{(2)}(1) - X_2^{(1)}(3)\right]^2}$$

$$+ \sqrt{\left[X_1^{(2)}(1) - X_1^{(2)}(3)\right]^2 + \left[X_2^{(2)}(1) - X_2^{(2)}(3)\right]^2}$$

$$+ \sqrt{\left[X_1^{(2)}(1) - X_1^{(3)}(3)\right]^2 + \left[X_2^{(2)}(1) - X_2^{(3)}(3)\right]^2}$$

$$+ \sqrt{\left[X_1^{(3)}(1) - X_1^{(1)}(3)\right]^2 + \left[X_2^{(3)}(1) - X_2^{(1)}(3)\right]^2}$$

$$+ \sqrt{\left[X_1^{(3)}(1) - X_1^{(2)}(3)\right]^2 + \left[X_2^{(3)}(1) - X_2^{(2)}(3)\right]^2}$$

$$+ \sqrt{\left[X_1^{(3)}(1) - X_1^{(3)}(3)\right]^2 + \left[X_2^{(3)}(1) - X_2^{(3)}(3)\right]^2}$$

$$= \sqrt{(2/3)^2 + (1/3)^2} + \sqrt{(2/3)^2 + (1/3)^2} + \sqrt{(1/3)^2}$$

$$+ \sqrt{(2/3)^2} + \sqrt{(2/3)^2 + (1)^2} + \sqrt{(2/3)^2 + (1/3)^2}$$

$$+ \sqrt{(1/3)^2} + \sqrt{(1)^2} + \sqrt{(1/3)^2} + \sqrt{(2/3)^2 + (1/3)^2}$$

$$= 6.18.$$

The complete matrix of distances obtained by applying the definition of distance between each pair of the sampled trajectories (see Equation (3.3)) is

$$d_{ml} = \begin{bmatrix} 0.00 & & & & & \\ 5.50 & 0.00 & & & & \\ 6.18 & 5.31 & 0.00 & & & \\ 6.89 & 6.18 & 6.57 & 0.00 & & \\ 6.18 & 5.31 & 5.41 & 5.50 & 0.00 & \\ 7.52 & 5.99 & 5.52 & 7.31 & 5.77 & 0.00 \end{bmatrix}.$$

To select the optimal set of trajectories it is necessary to consider all the possible combinations of four trajectories out of six and to evaluate the squared root of the sum of their squared distances, D. Table 3.7 shows the possible combinations for the present exercises. The optimal set of trajectories is $\mathbf{B}^*_1, \mathbf{B}^*_3, \mathbf{B}^*_4, \mathbf{B}^*_6$.

(b) Figure 3.9 shows the histograms of the values sampled for X_1 and X_2 when using the optimal trajectories scheme (bottom plot) and the original trajectories scheme (top plot), assuming that the original scheme would have taken the first four generated (i.e. $\mathbf{B}^*_1$, $\mathbf{B}^*_2$,

Table 3.7 Possible combinations of four trajectories out of six and corresponding values of the measure used to identify the optimal set

Trajectories $\{i,j,l,m\}$	D	Trajectories $\{i,j,l,m\}$	D
1-2-3-4	15.022	1-3-5-6	15.685
1-2-3-5	13.871	1-4-5-6	16.098
1-2-3-6	14.815	2-3-4-5	14.049
1-2-4-5	14.582	2-3-4-6	15.146
1-2-4-6	16.178	2-3-5-6	14.333
1-2-5-6	14.912	2-4-5-6	14.807
1-3-4-5	15.055	3-4-5-6	14.825
1-3-4-6	16.410		

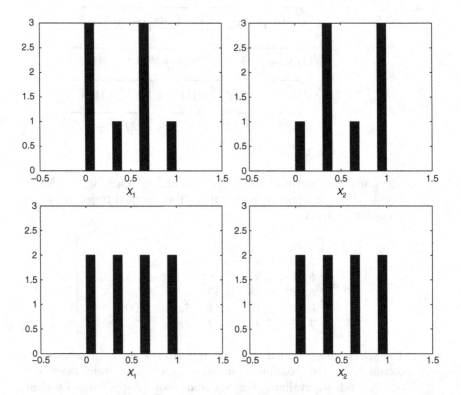

Figure 3.9 Original (top plots) versus optimized (bottom plots) sampled distributions obtained in Exercise 3

B^*_3 and B^*_4). Figure 3.10 plots the paths of the original (top plot) and optimized (bottom plot) trajectories. The optimized strategy has produced a better sample, in terms of both the level explored (the

SOLUTIONS 137

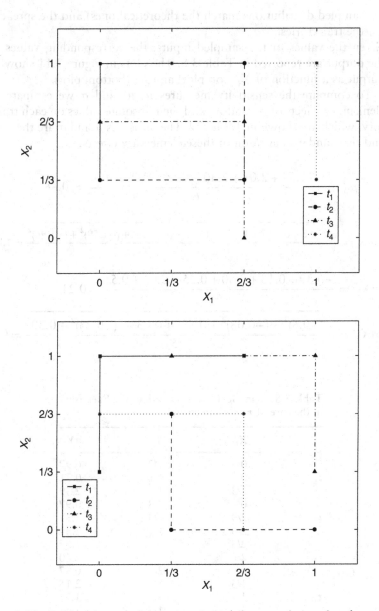

Figure 3.10 Original (top plot) versus optimized (bottom plot) paths of sampled trajectories for Exercise 3

sampled distributions match the theoretical ones) and the spread of the trajectories.

4. Given the values of the sampled inputs, the corresponding values for the output are reported in Table 3.8. The plots in Figure 3.11 show the output as a function of X_1 (top plot) and X_2 (bottom plot).

To compute the sensitivity measures μ, μ^* and σ we estimate the elementary effects of X_1 and X_2 and their absolute values on each trajectory, which are shown in Table 3.9. The measures μ and σ are the mean and the standard deviation of these elementary effects:

$$\mu(X_1) = \frac{2.52 + 2.01 + 2.30 - 0.66 - 0.93 - 1.30}{6} = 0.66$$

$$\sigma(X_1) = \sqrt{\frac{(2.52-0.66)^2 + 1.35^2 + 1.64^2 + (-1.32)^2 + (-1.59)^2 + (-1.96)^2}{5}} = 1.79$$

$$\mu(X_2) = \frac{-0.39 + 0.13 + 0.80 + 0.25 - 0.02 + 0.51}{6} = 0.21$$

$$\sigma(X_2) = \sqrt{\frac{(-0.6)^2 + (-0.08)^2 + 0.59^2 + 0.03^2 + (-0.23)^2 + 0.30^2}{5}} = 0.41$$

Table 3.8 Sample and output values for Exercise 4 in the case of $r = 6$ trajectories

	X_1	X_2	Y
	0	1/3	0.97
t_1	0	1	0.71
	2/3	1	2.39
	0	1/3	0.97
t_2	2/3	1/3	2.30
	2/3	1	2.39
	2/3	0	1.87
t_3	2/3	2/3	2.40
	0	2/3	0.87
	1/3	1	2.15
t_4	1	1	1.71
	1	1/3	1.54
	1/3	1	2.15
t_5	1/3	1/3	2.17
	1	1/3	1.54
	1/3	2/3	2.20
t_6	1/3	0	1.87
	1	0	1.00

SOLUTIONS

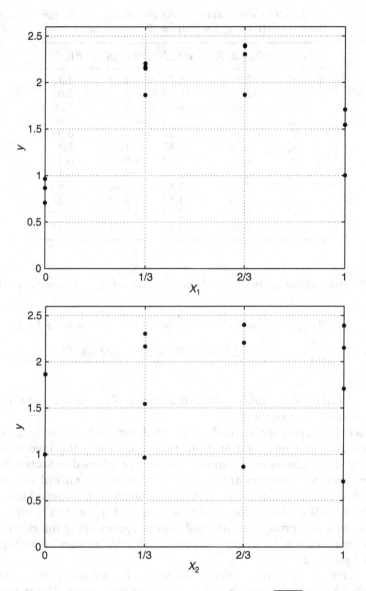

Figure 3.11 Scatterplots of $y = \sin(\pi X_1) + \cos(\pi X_2/4) + \sqrt{X_1 X_2}$ as a function of X_1 (top plot) and X_2 (bottom plot)

Table 3.9 Computation of the elementary effects and their absolute values for Exercise 4

		$y(X_i+\Delta, X_j)$	$y(X_i, X_j)$	$EE_i(X)$	$|EE_i(X)|$
X_1	t_1	2.39	0.71	2.52	2.52
	t_2	2.30	0.97	2.01	2.01
	t_3	2.40	0.87	2.30	2.30
	t_4	1.71	2.15	−0.66	0.66
	t_5	1.54	2.17	−0.93	0.93
	t_6	1.00	1.87	−1.30	1.3
X_2	t_1	0.71	0.97	−0.39	0.39
	t_2	2.39	2.30	0.13	0.13
	t_3	2.40	1.87	0.80	0.80
	t_4	1.71	1.54	0.25	0.25
	t_5	2.15	2.17	−0.02	0.02
	t_6	2.20	1.87	0.51	0.51

μ^* is instead the mean of the absolute values of these elementary effects:

$$\mu^*(X_1) = \frac{2.52+2.01+2.30+0.66+0.93+1.30}{6} = 1.62$$

$$\mu^*(X_2) = \frac{0.39+0.13+0.80+0.25+0.02+0.51}{6} = 0.35.$$

Both values of μ^* and σ show that factor X_1 is unequivocally more important than factor X_2.

5. Since the factors are normally distributed, we need to sample the quantiles and subsequently transform the values using the inverse of the normal cumulative distribution function, as explained in Section 3.4. In this exercise, quantiles are obtained by cutting the tails of the normal distribution at quantiles 5 and 95% and dividing this interval into three intervals. We leave to the reader the option of repeating the same exercise without cutting the tails, and instead apportioning the entire space of quantiles into four intervals and using the bin centres, as exemplified in Exercise 2.

Table 3.10 reports a sample composed of $r=4$ trajectories. The last column shows the corresponding values of the output. The trajectories are labelled as $t_i, i = 1, 2, 3, 4$.

Consider, for instance, factor X_1. Its elementary effects are estimated as shown in Table 3.11. Note that to estimate $EE_i(X)$ and its absolute value, the rescaling factor is $\Delta = 2/3$ (i.e. the variation of the factor in the space of the quantiles $[0, 1]$ is used). Table 3.12 reports the elementary effects for each trajectory and for all the factors.

Table 3.10 Sampled trajectories (labeled t_1, t_2, t_3, and t_4) and corresponding values for the Oakley–O'Hagan function

t_1

X_1	X_2	X_3	X_4	X_5	X_6	X_7	X_8	X_9	X_{10}	X_{11}	X_{12}	X_{13}	X_{14}	X_{15}	Y
1.64	−1.64	−1.64	0.39	−0.39	0.39	−1.64	−1.64	−0.39	−0.39	1.64	1.64	−0.39	0.39	1.64	24.90
1.64	−1.64	−1.64	0.39	−0.39	−1.64	−1.64	−1.64	−0.39	−0.39	1.64	1.64	−0.39	0.39	1.64	22.72
1.64	−1.64	−1.64	0.39	−0.39	−1.64	−1.64	−1.64	1.64	−0.39	1.64	1.64	−0.39	0.39	1.64	21.04
1.64	−1.64	−1.64	0.39	−0.39	−1.64	−1.64	0.39	1.64	−0.39	1.64	1.64	−0.39	0.39	1.64	16.01
1.64	−1.64	−1.64	−1.64	−0.39	−1.64	−1.64	0.39	1.64	−0.39	1.64	1.64	−0.39	0.39	1.64	10.40
1.64	−1.64	−1.64	−1.64	−0.39	−1.64	−1.64	0.39	1.64	−0.39	1.64	−0.39	−0.39	0.39	1.64	10.04
1.64	−1.64	−1.64	−1.64	−0.39	−1.64	−1.64	0.39	1.64	−0.39	1.64	−0.39	−0.39	−1.64	1.64	8.60
1.64	0.39	−1.64	−1.64	−0.39	−1.64	−1.64	0.39	1.64	−0.39	1.64	−0.39	−0.39	−1.64	1.64	13.39
1.64	0.39	−1.64	−1.64	−0.39	−1.64	−1.64	0.39	1.64	−0.39	1.64	−0.39	−0.39	−1.64	−0.39	4.69
1.64	0.39	−1.64	−1.64	−0.39	−1.64	−1.64	0.39	1.64	1.64	1.64	−0.39	−0.39	−1.64	−0.39	8.02
1.64	0.39	−1.64	−1.64	1.64	−1.64	−1.64	0.39	1.64	1.64	1.64	−0.39	−0.39	−1.64	−0.39	9.98
1.64	0.39	−1.64	−1.64	1.64	−1.64	−1.64	0.39	1.64	1.64	−0.39	−0.39	−0.39	−1.64	−0.39	3.75
1.64	0.39	−1.64	−1.64	1.64	−1.64	0.39	0.39	1.64	1.64	−0.39	−0.39	−0.39	−1.64	−0.39	1.33
1.64	0.39	0.39	−1.64	1.64	−1.64	0.39	0.39	1.64	1.64	−0.39	−0.39	−0.39	−1.64	−0.39	2.59
−0.39	0.39	0.39	−1.64	1.64	−1.64	0.39	0.39	1.64	1.64	−0.39	−0.39	−0.39	−1.64	−0.39	6.37
−0.39	0.39	0.39	−1.64	1.64	−1.64	0.39	0.39	1.64	1.64	−0.39	−0.39	1.64	−1.64	−0.39	9.99

Table 3.10 (Continued)

								t_2							
X_1	X_2	X_3	X_4	X_5	X_6	X_7	X_8	X_9	X_{10}	X_{11}	X_{12}	X_{13}	X_{14}	X_{15}	Y
−0.39	−0.39	0.39	0.39	−0.39	−0.39	−0.39	−1.64	0.39	−1.64	0.39	−0.39	−0.39	1.64	1.64	14.04
−0.39	−0.39	0.39	0.39	−0.39	−0.39	−0.39	−1.64	0.39	−1.64	0.39	−0.39	1.64	1.64	1.64	12.86
−0.39	−0.39	0.39	0.39	−0.39	−0.39	1.64	−1.64	0.39	−1.64	0.39	−0.39	1.64	1.64	1.64	15.42
−0.39	−0.39	0.39	0.39	−0.39	−0.39	1.64	−1.64	−1.64	−1.64	0.39	−0.39	1.64	1.64	1.64	14.56
−0.39	1.64	0.39	0.39	−0.39	−0.39	1.64	−1.64	−1.64	−1.64	0.39	−0.39	1.64	1.64	1.64	6.17
−0.39	1.64	0.39	0.39	1.64	−0.39	1.64	−1.64	−1.64	−1.64	0.39	−0.39	1.64	1.64	1.64	0.96
−0.39	1.64	0.39	0.39	1.64	−0.39	1.64	−1.64	−1.64	−1.64	0.39	1.64	1.64	1.64	1.64	2.91
−0.39	1.64	−1.64	0.39	1.64	−0.39	1.64	−1.64	−1.64	−1.64	0.39	1.64	1.64	1.64	1.64	−3.65
−0.39	1.64	−1.64	0.39	1.64	1.64	1.64	−1.64	−1.64	−1.64	0.39	1.64	1.64	1.64	1.64	−6.31
−0.39	1.64	−1.64	0.39	1.64	1.64	1.64	−1.64	−1.64	−1.64	0.39	1.64	1.64	1.64	−0.39	−0.70
−0.39	1.64	−1.64	0.39	1.64	1.64	1.64	−1.64	−1.64	−1.64	−1.64	1.64	1.64	1.64	−0.39	−11.79
−0.39	1.64	−1.64	0.39	1.64	1.64	1.64	−1.64	−1.64	−1.64	−1.64	1.64	1.64	−0.39	−0.39	−16.95
−0.39	1.64	−1.64	0.39	1.64	1.64	1.64	−1.64	−1.64	0.39	−1.64	1.64	1.64	−0.39	−0.39	−15.54
−0.39	1.64	−1.64	0.39	1.64	1.64	1.64	0.39	−1.64	0.39	−1.64	1.64	1.64	−0.39	−0.39	−3.01
−0.39	1.64	−1.64	−1.64	1.64	1.64	1.64	0.39	−1.64	0.39	−1.64	1.64	1.64	−0.39	−0.39	−2.12
1.64	1.64	−1.64	−1.64	1.64	1.64	1.64	0.39	−1.64	0.39	−1.64	1.64	1.64	−0.39	−0.39	−14.36

							t_3									
X_1	X_2	X_3	X_4	X_5	X_6	X_7	X_8	X_9	X_{10}	X_{11}	X_{12}	X_{13}	X_{14}	X_{15}	Y
0.39	0.39	−0.39	−0.39	0.39	0.39	−1.64	0.39	−0.39	1.64	−0.39	0.39	−1.64	1.64	−1.64	3.96
0.39	0.39	−0.39	−0.39	0.39	0.39	−1.64	0.39	−0.39	1.64	1.64	0.39	−1.64	1.64	−1.64	2.92
0.39	0.39	−0.39	1.64	0.39	0.39	−1.64	0.39	−0.39	1.64	1.64	0.39	−1.64	1.64	−1.64	9.78
0.39	0.39	−0.39	1.64	0.39	0.39	−1.64	0.39	−0.39	1.64	1.64	−1.64	−1.64	1.64	−1.64	1.96
−1.64	0.39	−0.39	1.64	0.39	0.39	−1.64	0.39	−0.39	1.64	1.64	−1.64	−1.64	1.64	−1.64	−5.53
−1.64	0.39	−0.39	1.64	0.39	0.39	0.39	0.39	−0.39	1.64	1.64	−1.64	−1.64	1.64	−1.64	1.87
−1.64	0.39	1.64	1.64	0.39	0.39	0.39	0.39	−0.39	1.64	1.64	−1.64	−1.64	1.64	−1.64	−0.45
−1.64	−1.64	1.64	1.64	0.39	0.39	0.39	0.39	−0.39	1.64	1.64	−1.64	−1.64	1.64	−1.64	−3.81
−1.64	−1.64	1.64	1.64	0.39	0.39	0.39	−1.64	−0.39	1.64	1.64	−1.64	−1.64	1.64	−1.64	0.78
−1.64	−1.64	1.64	1.64	−1.64	0.39	0.39	−1.64	−0.39	1.64	1.64	−1.64	−1.64	1.64	−1.64	−0.27
−1.64	−1.64	1.64	1.64	−1.64	0.39	0.39	−1.64	1.64	1.64	1.64	−1.64	−1.64	1.64	−1.64	0.17
−1.64	−1.64	1.64	1.64	−1.64	−1.64	0.39	−1.64	1.64	1.64	1.64	−1.64	−1.64	1.64	−1.64	−5.89
−1.64	−1.64	1.64	1.64	−1.64	−1.64	0.39	−1.64	1.64	1.64	1.64	−1.64	0.39	1.64	−1.64	−0.63
−1.64	−1.64	1.64	1.64	−1.64	−1.64	0.39	−1.64	1.64	1.64	1.64	−1.64	0.39	1.64	0.39	8.37
−1.64	−1.64	1.64	1.64	−1.64	−1.64	0.39	−1.64	1.64	1.64	1.64	−1.64	0.39	−0.39	0.39	5.65
−1.64	−1.64	1.64	1.64	−1.64	−1.64	0.39	−1.64	1.64	−0.39	1.64	−1.64	0.39	−0.39	0.39	−1.91

Table 3.10 (Continued)

t_4

X_1	X_2	X_3	X_4	X_5	X_6	X_7	X_8	X_9	X_{10}	X_{11}	X_{12}	X_{13}	X_{14}	X_{15}	Y
1.64	−0.39	1.64	−0.39	0.39	0.39	−1.64	1.64	−1.64	−1.64	0.39	0.39	−1.64	0.39	0.39	5.05
1.64	−0.39	1.64	−0.39	−1.64	0.39	−1.64	1.64	−1.64	−1.64	0.39	0.39	−1.64	0.39	0.39	3.91
1.64	−0.39	1.64	1.64	−1.64	0.39	−1.64	1.64	−1.64	−1.64	0.39	0.39	−1.64	0.39	0.39	6.81
1.64	−0.39	1.64	1.64	−1.64	0.39	−1.64	1.64	−1.64	0.39	0.39	0.39	−1.64	0.39	0.39	5.58
1.64	1.64	1.64	1.64	−1.64	0.39	−1.64	1.64	−1.64	0.39	0.39	0.39	−1.64	0.39	0.39	11.79
1.64	1.64	1.64	1.64	−1.64	0.39	−1.64	1.64	−1.64	0.39	0.39	0.39	−1.64	0.39	−1.64	5.00
1.64	1.64	1.64	1.64	−1.64	0.39	−1.64	1.64	−1.64	0.39	−1.64	0.39	−1.64	0.39	−1.64	0.77
1.64	1.64	1.64	1.64	−1.64	0.39	−1.64	1.64	−1.64	0.39	−1.64	0.39	−1.64	0.39	−1.64	1.38
−0.39	1.64	1.64	1.64	−1.64	0.39	−1.64	1.64	−1.64	0.39	−1.64	0.39	−1.64	−1.64	−1.64	−10.16
−0.39	1.64	1.64	1.64	−1.64	0.39	−1.64	1.64	−1.64	0.39	−1.64	0.39	−1.64	−1.64	−1.64	3.51
−0.39	1.64	1.64	1.64	−1.64	0.39	−1.64	1.64	0.39	0.39	−1.64	0.39	0.39	−1.64	−1.64	10.29
−0.39	1.64	1.64	1.64	−1.64	0.39	−1.64	−0.39	0.39	0.39	−1.64	0.39	0.39	−1.64	−1.64	−0.14
−0.39	1.64	1.64	1.64	−1.64	0.39	−1.64	−0.39	0.39	0.39	−1.64	−1.64	0.39	−1.64	−1.64	−9.34
−0.39	1.64	−0.39	1.64	−1.64	0.39	−1.64	−0.39	0.39	0.39	−1.64	−1.64	0.39	−1.64	−1.64	−7.57
−0.39	1.64	−0.39	1.64	−1.64	0.39	0.39	−0.39	0.39	0.39	−1.64	−1.64	0.39	−1.64	−1.64	−7.90
−0.39	1.64	−0.39	1.64	−1.64	−1.64	0.39	−0.39	0.39	0.39	−1.64	−1.64	0.39	−1.64	−1.64	−7.06

SOLUTIONS

Table 3.11 Estimation of the elementary effects for factor X_1 of Exercise 5

	$y(X_1+\Delta, X_{\sim 1})$	$y(X_1, X_{\sim 1})$	$EE_1(X)$	$|EF_1(X)|$
t_1	2.59	6.37	−5.67	5.67
t_2	−14.36	−2.12	−18.36	18.36
t_3	1.96	−5.53	11.24	11.24
t_4	0.77	1.38	−0.92	0.92

Table 3.12 Elementary effects for all the factors of Exercise 5 using the $r = 4$ trajectories shown in Table 3.10

	t_1	t_2	t_3	t_4
X_1	−5.67	−18.36	11.24	−0.92
X_2	7.18	−12.58	5.04	9.31
X_3	1.89	9.84	−3.47	−2.66
X_4	8.42	−1.33	10.29	4.34
X_5	2.93	−7.82	1.57	1.71
X_6	3.28	−3.99	9.10	−1.25
X_7	−3.62	3.85	11.1	−0.5
X_8	−7.55	18.8	−6.89	15.65
X_9	−2.51	1.30	0.66	10.18
X_{10}	5.00	2.11	11.33	−1.83
X_{11}	9.34	16.64	−1.56	6.35
X_{12}	0.54	2.93	11.72	13.81
X_{13}	5.43	−1.78	7.90	20.5
X_{14}	2.15	7.74	4.08	17.31
X_{15}	13.05	−8.41	13.49	10.19

For instance, the three sensitivity measures for factor X_1 are obtained as follows:

$$\mu(X_1) = \frac{-5.67 - 18.36 + 11.24 - 0.92}{4} = -3.42$$

$$\mu^*(X_1) = \frac{5.67 + 18.36 + 11.24 + 0.92}{4} = 9.05$$

$$\sigma(X_1) = \sqrt{\frac{(-2.24)^2 + (-14.93)^2 + (14.67)^2 + (2.51)^2}{3}} = 12.24.$$

Columns 2–4 of Table 3.13 show the sensitivity analysis results for the Oakley–O'Hagan function for $r = 4$. Results for $r = 10$ are also reported in the table. The last two columns show, for comparative purposes, the analytical values of the total- and first-order sensitivity indices for the factors.

Table 3.13 Sensitivity analysis results for the Oakley–O'Hagan function of Exercise 5 using $r = 4$ and $r = 10$ trajectories. The last two columns report the analytical values of the total- and first-order sensitivity indices

	$r = 4$			$r = 10$			Analytics	Analytics
	μ^*	μ	σ	μ^*	μ	σ	S_T	S
X_1	9.05	−3.42	12.24	6.21	0.98	6.77	0.059	0.002
X_2	8.53	2.24	10.03	3.35	1.28	4.46	0.063	0.000
X_3	4.46	1.40	6.10	6.54	0.77	7.59	0.036	0.001
X_4	6.09	5.43	5.14	4.54	0.63	5.49	0.055	0.003
X_5	3.51	−0.4	4.98	3.13	−0.67	4.51	0.024	0.003
X_6	4.40	1.78	5.72	5.93	3.58	7.76	0.041	0.023
X_7	4.77	2.71	6.38	7.18	4.72	8.36	0.058	0.024
X_8	12.22	5.00	14.17	5.26	2.38	6.30	0.082	0.027
X_9	3.66	2.41	5.44	8.25	7.10	7.87	0.097	0.046
X_{10}	5.07	4.15	5.54	2.79	1.04	3.91	0.036	0.015
X_{11}	8.47	7.69	7.53	8.82	8.82	5.82	0.151	0.102
X_{12}	7.25	7.25	6.50	8.33	8.33	4.12	0.148	0.136
X_{13}	8.91	8.02	9.28	10.65	7.95	9.64	0.142	0.102
X_{14}	7.82	7.82	6.74	7.10	5.32	6.71	0.141	0.105
X_{15}	11.29	7.08	10.43	7.83	7.83	4.01	0.155	0.123

In our experiments, the use of $r = 4$ trajectories has not helped to discern the relative importance of the factors; for instance, the most important factor according to μ^* is X_8, while according to the analytic values of ST it is only the seventh most important factor. These problems with the Oakley–O'Hagan function are mainly due to the fact that interactions between all pairs of factors play a relevant role in the model. Increasing the number of trajectories to $r = 10$ allows us to better understand the sensitivity of the model. Figure 3.12 shows in a barplot the value of μ^* for the inputs of the Oakley–O'Hagan function. This plot highlights that in this model it is not possible to identify a subset of factors to be fixed without affecting the output. The high values of σ confirm that this is related to the presence of non-negligible interactions even for factors with low main effects $(X_1, .., X_{10})$.

6. According to the matrix **G** and considering the results of the previous exercise, factors are grouped so that all inputs with a high main effect are in the first group. Table 3.14 reports the trajectories obtained for the 15 factors grouped according to the matrix **G**. As in Exercise 5, since factors are normally distributed quantiles are sampled first and then levels are obtained using the inverse cumulative distribution function.

Table 3.15 shows the absolute values of the elementary effects estimated for each group using the 10 trajectories. The value of μ^* for the

SOLUTIONS

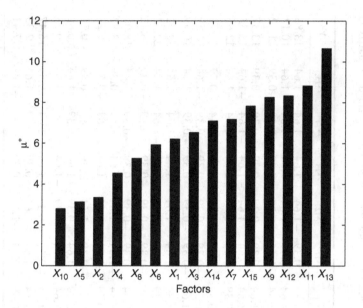

Figure 3.12 Barplot of μ^* for the Oakley–O'Hagan function of Exercise 5, using $r = 10$ trajectories

group (last column) is the average over the 10 estimates. It is clearly verified that the first group is more important than the second. However, it is also confirmed that the factors in G_2 are actually playing a role on the output (see also the values of their S_T in Table 3.13) and thus cannot be discarded.

7. Since the number of factors is quite high, we do not report the complete sample used for the sensitivity analysis. However, Figure 3.13 shows the histogram of y for a design with $p = 4$ levels and $r = 10$ trajectories (i.e. $N = 210$ model evaluations). Table 3.16 reports the sensitivity analysis results. Figure 3.14 helps in reading these results by presenting the scatterplot $\{\mu^*, \sigma\}$. This is the representation suggested by Morris (1991).

Figure 3.15 presents a horizontal barplot of the values of μ^* for the inputs of the function y. Factors $X_{11}, \ldots, X_{20}$ are seen at a glance to be negligible. In fact, looking at the structure of y, their coefficients are never large. Factors X_8, X_9, X_{10} have a strong linear effect (for factors 1–10, $\beta_i = 20$). The linear plus nonlinear influence of X_7, due to the transformation $\omega_i = 2\left(1.1 X_i/(X_i + 0.1) - 1/2\right)$, is also captured by the high value of the standard deviation. Finally, factors $X_1, \ldots, X_6$ are all identified as influential and to have linear, nonlinear and nonadditive effects.

Table 3.14 Trajectories obtained for Exercise 6 (Oakley-O'Hagan function, $r = 10$) in case the two groups described by matrix G are considered

X_1	X_2	X_3	X_4	X_5	X_6	X_7	X_8	X_9	X_{10}	X_{11}	X_{12}	X_{13}	X_{14}	X_{15}	Y
0.39	−0.39	−1.64	0.39	−0.39	−0.39	0.39	0.39	−1.64	−0.39	0.39	−1.64	1.64	1.64	1.64	13.85
−1.64	1.64	0.39	−1.64	1.64	1.64	−1.64	−1.64	−1.64	1.64	0.39	−1.64	1.64	1.64	1.64	−10.11
−1.64	1.64	0.39	−1.64	1.64	1.64	−1.64	−1.64	0.39	1.64	−1.64	0.39	−0.39	−0.39	−0.39	1.12
−0.39	−0.39	0.39	−1.64	0.39	−1.64	1.64	1.64	0.39	−1.64	0.39	−0.39	0.39	0.39	0.39	13.37
−0.39	−0.39	0.39	−1.64	0.39	−1.64	1.64	1.64	−1.64	−1.64	−1.64	1.64	−1.64	−1.64	−1.64	−13.33
1.64	1.64	−1.64	0.39	−1.64	0.39	−0.39	−0.39	−1.64	0.39	−1.64	1.64	−1.64	−1.64	−1.64	−12.68
1.64	−1.64	1.64	1.64	1.64	−0.39	−1.64	−1.64	−0.39	−0.39	1.64	1.64	−1.64	1.64	0.39	15.65
1.64	−1.64	1.64	1.64	1.64	−0.39	−1.64	−1.64	1.64	−0.39	−0.39	−0.39	0.39	−0.39	−1.64	1.59
−0.39	−0.39	−0.39	−0.39	−0.39	1.64	0.39	0.39	1.64	1.64	−0.39	−0.39	0.39	−0.39	−1.64	10.99
−0.39	−1.64	−1.64	0.39	0.39	−0.39	0.39	−0.39	−0.39	1.64	1.64	1.64	1.64	−1.64	1.64	22.92
1.64	0.39	0.39	−1.64	−1.64	1.64	−1.64	1.64	−0.39	−0.39	1.64	1.64	1.64	1.64	1.64	22.62
1.64	0.39	0.39	−1.64	−1.64	1.64	−1.64	1.64	1.64	−0.39	−0.39	−0.39	−0.39	0.39	−0.39	6.42
1.64	1.64	−1.64	1.64	1.64	−0.39	−0.39	1.64	0.39	1.64	0.39	−0.39	0.39	−0.39	1.64	18.01
1.64	1.64	−1.64	1.64	1.64	−0.39	−0.39	1.64	−1.64	1.64	−1.64	1.64	−1.64	1.64	−0.39	3.82
−0.39	−0.39	0.39	−0.39	−0.39	1.64	1.64	−0.39	−1.64	−0.39	−1.64	1.64	−1.64	1.64	−0.39	9.27
0.39	−1.64	−1.64	−0.39	−1.64	−1.64	−0.39	−1.64	−0.39	−0.39	−0.39	−1.64	0.39	−1.64	0.39	−0.71
0.39	−1.64	−1.64	−0.39	−1.64	−1.64	−0.39	−1.64	1.64	−0.39	1.64	0.39	−1.64	0.39	−1.64	0.62
−1.64	−0.39	−0.39	1.64	0.39	0.39	1.64	0.39	1.64	1.64	1.64	0.39	−1.64	0.39	−1.64	10.63
−1.64	−1.64	0.39	−1.64	1.64	1.64	1.64	1.64	1.64	−1.64	−0.39	0.39	−1.64	1.64	0.39	16.47
−1.64	−1.64	0.39	−1.64	1.64	1.64	1.64	1.64	−0.39	−1.64	1.64	−1.64	0.39	−0.39	−1.64	2.33
0.39	0.39	−1.64	0.39	−0.39	−0.39	−0.39	−0.39	−0.39	0.39	1.64	−1.64	0.39	−0.39	−1.64	7.41

−1.64	1.64	0.39	−0.39	1.64	1.64	1.64	1.64	−0.39	−0.39	−0.39
−1.64	1.64	0.39	−0.39	1.64	1.64	1.64	−0.39	1.64	1.64	1.64
0.39	−0.39	−1.64	1.64	0.39	−0.39	−0.39	−0.39	1.64	1.64	1.64
−1.64	−0.39	−0.39	−0.39	1.64	1.64	1.64	0.39	1.64	−1.64	−1.64
0.39	1.64	−0.39	1.64	0.39	−1.64	−0.39	0.39	−1.64	−1.64	1.64
0.39	1.64	−0.39	−1.64	−1.64	−1.64	0.39	−0.39	−1.64	0.39	1.64
−0.39	−1.64	−1.64	−1.64	1.64	0.39	−1.64	−0.39	0.39	−1.64	−1.64
−1.64	0.39	0.39	0.39	0.39	−0.39	−0.39	0.39	−0.39	1.64	1.64
0.39	−1.64	−1.64	1.64	−1.64	−1.64	−0.39	−1.64	−0.39	1.64	1.64
0.39	−1.64	−1.64	1.64	−1.64	1.64	−1.64	1.64	1.64	1.64	0.39

9.8	20.97	12.8	−0.45	−5.59	−2.07	2.29	−6.87	17.23

Table 3.15 Absolute values of the elementary effects obtained for the two groups in Exercise 6. μ^* is the average over the 10 estimates

	t_1	t_2	t_3	t_4	t_5	t_6	t_7	t_8	t_9	t_{10}	μ^*
G_1	16.86	40.06	21.09	24.3	21.29	1.99	21.2	16.75	5.28	36.16	20.5
G_2	35.95	0.97	14.1	0.45	8.18	15.01	7.62	12.26	7.70	13.75	11.6

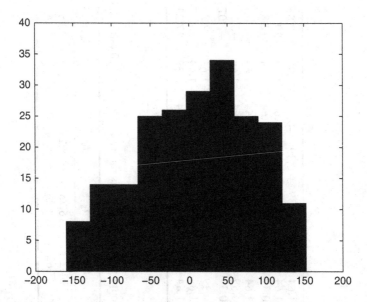

Figure 3.13 Empirical density of the Morris function obtained for Exercise 7 with $r = 10$ trajectories

Table 3.16 Sensitivity analysis for the Morris function of Exercise 7 using $r = 10$ trajectories

	μ^*	μ	σ
X_1	71.644	55.256	54.648
X_2	64.290	64.29	37.737
X_3	88.207	55.572	94.737
X_4	73.038	9.310	91.918
X_5	44.777	32.874	55.793
X_6	45.280	35.187	44.940
X_7	27.039	27.039	23.594
X_8	40.716	40.716	5.789
X_9	40.385	40.385	6.101
X_{10}	39.706	39.706	6.400
X_{11}	5.252	1.232	6.608
X_{12}	4.324	1.494	4.990
X_{13}	6.268	0.922	7.892
X_{14}	5.183	0.192	6.845
X_{15}	8.842	3.797	9.773
X_{16}	6.201	2.183	7.511
X_{17}	3.389	−1.335	4.291
X_{18}	4.903	−1.540	5.338
X_{19}	4.989	−0.766	5.686
X_{20}	3.854	−2.407	5.087

SOLUTIONS

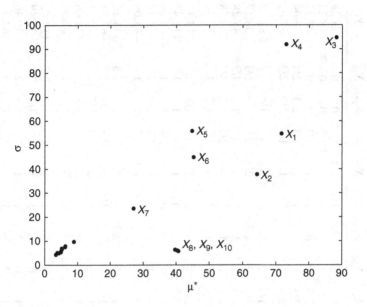

Figure 3.14 Sensitivity analysis results for the Morris function of Exercise 7 using $r = 10$ trajectories

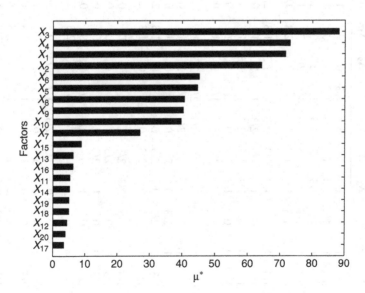

Figure 3.15 Horizontal barplot of the value of μ^* for the the Morris function of Exercise 7, obtained using $r = 10$ trajectories

Table 3.17 Sample of $r = 10$ trajectories for the analysis by group of Exercise 8

	X_1	X_2	X_3	X_4	X_5	X_6	X_7	X_8	X_9	X_{10}	X_{11}	X_{12}	X_{13}	X_{14}	X_{15}	X_{16}	X_{17}	X_{18}	X_{19}	X_{20}	Y
t_1	0	0	1	1	1/3	1	1	1/3	0	0	1	0	0	1/3	2/3	1/3	1/3	1	0	0	58.28
	0	0	1	1	1/3	1	1	1/3	0	0	1	0	0	1	0	1	1	1	0	0	68.24
	0	0	1	1	1/3	1/3	1/3	1	2/3	0	1	0	2/3	1	0	1	1	1/3	0	0	87.75
	0	0	1	1/3	1	1/3	1/3	1	2/3	0	1	0	2/3	1	0	1	1	1/3	2/3	2/3	23.04
	2/3	2/3	1/3	1/3	1	1/3	1/3	1	2/3	2/3	1/3	2/3	2/3	1	0	1	1	1/3	2/3	2/3	63.61
t_2	1	1/3	2/3	0	1/3	1/3	2/3	1	1/3	1/3	2/3	1/3	1/3	0	1/3	0	2/3	0	1	1/3	81.50
	1/3	1	0	0	1/3	1/3	2/3	1	1/3	1	0	1	1/3	0	1/3	0	2/3	0	1	1/3	38.17
	1/3	1	0	0	1/3	1/3	2/3	1	1/3	1	0	1	1/3	2/3	1	2/3	0	0	1	1/3	47.28
	1/3	1	0	0	1/3	1	0	1/3	1	1	0	1	1	2/3	1	2/3	0	2/3	1	1/3	43.58
	1/3	1	0	2/3	1	1	0	1/3	1	1	0	1	1	2/3	1	2/3	0	2/3	1/3	1	72.56
t_3	1	1	1/3	0	2/3	0	2/3	2/3	2/3	1	0	0	2/3	1	1/3	0	1	0	1/3	1/3	115.15
	1	1	1/3	2/3	0	0	2/3	2/3	2/3	1	0	0	2/3	1	1/3	0	1	0	1	1	116.45
	1	1/3	1/3	2/3	0	2/3	0	0	0	1	0	0	0	1	1/3	0	0	2/3	1	1	9.06
	1/3	1/3	1	2/3	0	2/3	0	0	0	1/3	2/3	2/3	0	1	1/3	0	1	2/3	1	1	−50.12
	1/3	1/3	1	2/3	0	2/3	0	0	0	1/3	2/3	2/3	0	1/3	1	2/3	1/3	2/3	1/3	1	−48.06
t_4	0	1/3	0	2/3	1/3	2/3	1	0	1	0	0	1	1/3	0	2/3	2/3	2/3	2/3	1	1/3	−33.16
	0	1/3	0	0	1	2/3	1	0	1	0	0	1	1/3	0	2/3	2/3	2/3	2/3	1/3	1	−67.02
	0	1/3	0	0	1	2/3	1	1	1	0	0	1	1/3	2/3	0	0	0	2/3	1/3	1	−60.14
	0	1/3	0	0	1	0	1/3	2/3	1/3	0	0	1	1	2/3	0	0	0	0	1/3	1	−132.66
	2/3	1	2/3	0	1	0	1/3	2/3	1/3	2/3	2/3	1/3	1	2/3	1	1/3	1/3	0	1/3	1	81.88
t_5	1/3	0	2/3	0	0	1/3	2/3	1/3	1/3	0	1/3	2/3	1/3	0	2/3	1	1	1/3	1/3	2/3	−118.82
	1/3	0	2/3	0	0	1	0	1	1	0	1/3	2/3	1	0	2/3	1	1	1	1/3	2/3	−32.49
	1	2/3	0	2/3	0	1	0	1	1	2/3	1	0	0	0	2/3	1	1	1	1/3	2/3	30.46
	1	2/3	0	2/3	2/3	1	0	1	1	2/3	1	0	1	0	2/3	1	1	1	1	0	62.69
	1	2/3	0	2/3	2/3	1	0	1	1	2/3	1	0	1	2/3	0	1/3	1/3	1	1	0	57.40

	1/3	2/3	0	1/3	0	1	0	1/3	1/3	2/3	1/3	2/3	1	0	2/3	2/3	−49.84
	1/3	2/3	0	1	0	1	0	1/3	1/3	2/3	1/3	2/3	1	0	0	0	36.19
t_6	1	0	2/3	1	0	1	0	1	1/3	2/3	1/3	2/3	1	0	0	0	86.51
	1	0	2/3	1	2/3	1/3	2/3	1	1	2/3	1/3	2/3	1	2/3	2/3	0	104.25
	1	0	2/3	1	2/3	1/3	2/3	1	1	0	1	0	1/3	2/3	0	0	97.56
	0	1	2/3	1	1	1	0	0	1/3	1	1	1	2/3	1	2/3	1	−13.28
	0	1	2/3	1	1	1	0	0	1/3	1	1	1	0	1	2/3	1	−17.35
t_7	0	1	2/3	1/3	1	1/3	2/3	0	1/3	1	1	1	0	1	1/3	1/3	25.31
	2/3	1/3	0	1/3	1	1/3	2/3	0	1	1	1	1/3	0	1	1/3	1/3	1.93
	2/3	1/3	0	1/3	1	1	2/3	2/3	1	1	1	1/3	0	0	1/3	1/3	38.27
	1/3	1	1/3	0	0	1/3	2/3	2/3	2/3	2/3	1	0	0	1/3	1/3	1/3	−103.13
	1	1	1	2/3	0	1/3	2/3	2/3	2/3	0	1	0	0	1/3	1/3	1/3	69.38
t_8	1	1	1	2/3	1	1/3	2/3	2/3	2/3	0	1/3	2/3	2/3	1/3	1/3	1/3	34.50
	1	1	1	2/3	1	1	2/3	0	0	1/3	2/3	2/3	2/3	1/3	1/3	1/3	32.28
	1	1	2/3	2/3	1	1	2/3	0	0	1/3	2/3	2/3	2/3	1	1/3	1	−153.14
	1/3	1	1/3	1	1	2/3	1	1	2/3	1/3	1/3	2/3	1	0	0	1	86.29
	1/3	1	1/3	1	1	2/3	1	1	2/3	1/3	1/3	2/3	1	0	0	1	68.73
t_9	1/3	1	1/3	1	1	2/3	1	1	2/3	1/3	1/3	0	1	2/3	0	1	60.16
	1	1/3	1	1	1	1/3	2/3	1	0	1	1/3	0	0	2/3	0	1/3	21.83
	1	1/3	1	1	1	1/3	2/3	1	0	1	1/3	0	0	2/3	1/3	1/3	43.05
	1/3	1/3	1/3	1/3	1/3	1	2/3	0	2/3	1	2/3	2/3	0	1	1/3	2/3	38.96
	1/3	1/3	1/3	1	1	1	2/3	0	2/3	1	2/3	2/3	0	1	1/3	0	67.35
t_{10}	1/3	1/3	1/3	1	1	1/3	2/3	2/3	2/3	1	2/3	2/3	1/3	1/3	1/3	0	−7.34
	1	1	1	1	1	1/3	2/3	2/3	2/3	1	2/3	2/3	1/3	1/3	1	0	−205.31
	1	1	1	1	1	1/3	2/3	2/3	0	1/3	0	2/3	1/3	1	0	0	−212.21

Table 3.18 Absolute values of the elementary effects for each trajectory and each group of Exercise 8. The measure μ^* (last row) is the average over the 10 runs

	G_1	G_2	G_3	G_4
t_1	60.87	97.07	14.94	29.27
t_2	64.99	43.47	13.67	5.55
t_3	88.76	1.95	3.09	161.09
t_4	321.81	50.79	10.31	108.78
t_5	94.42	48.35	7.94	129.50
t_6	75.48	129.04	10.04	26.60
t_7	35.07	63.99	6.11	54.51
t_8	258.76	278.13	3.33	52.32
t_9	57.49	31.83	12.86	26.34
t_{10}	296.95	42.58	10.36	112.04
μ^*	135.46	78.72	9.26	70.60

8. Table 3.17 shows the sample for the groups and the corresponding values for the output. Note that slightly different results might be obtained since the parameters $\beta_i, \beta_{i,j}, \beta_{i,j,k}, \beta_{i,j,k,l}$ are all randomly generated. Table 3.18 reports the absolute values of the elementary effects, $|EE_i(\mathbf{X})|$, for each group along each of the 10 trajectories. Results confirm that there are three relevant groups (G_1, G_2 and G_4) and one group with low influence on the output (G_3). This is indeed correct since the factors in G_3 (X_{14}, X_{15}, X_{16} and X_{17}) have low main effects and low interactions (see also Table 3.16 above). Even the relative importance of G_1, G_2 and G_3 reflects the structure of the function y and the sensitivity analysis results of the previous exercise.

4

Variance-based Methods

HOW IS VARIANCE DECOMPOSITION RELATED TO SENSITIVITY ANALYSIS? WHEN IS IT WORTH USING VARIANCE-BASED TECHNIQUES INSTEAD OF SOMETHING ELSE? WHICH MEASURES CAN WE OBTAIN WITH VARIANCE-BASED METHODS? HOW ARE THESE MEASURES CALCULATED?

In this chapter we describe in more detail the variance-based methods that were introduced in Chapter 1. We first illustrate the settings that can be useful when dealing with modelling under conditions of uncertainty. We discuss the importance of a proper sensitivity test for a given setting. We sketch the historical background of variance-based methods, and discuss the properties of variance decompositions, from model independence to the capacity to evaluate the importance of groups of factors. Total effect indices are also introduced as a means of dealing synthetically with model complexity. We then illustrate two basic methods of computing the sensitivity indices, the Monte Carlo based design developed by Saltelli (2002) as well as the Random Balance Designs based on the Fourier Amplitude Sensitivity Test (FAST-RBD (Tarantola *et al.*, 2006)). Finally, we offer some examples.

4.1 DIFFERENT TESTS FOR DIFFERENT SETTINGS

It is common to find cases in the literature in which different sensitivity tests are applied to the same problem in a nonstructured fashion. This practice can yield a variety of results – e.g. in terms of ranking the factors in order of

---

Global Sensitivity Analysis. The Primer A. Saltelli, M. Ratto, T. Andres, F. Campolongo, J. Cariboni, D. Gatelli, M. Saisana and S. Tarantola © 2008 John Wiley & Sons, Ltd

importance – with no guidance as to which we should believe or privilege. As discussed in Chapter 1, we suggest instead a careful consideration of (a) the output of interest and (b) the concept of 'importance', as it applies to the problem at hand. This would in general allow for the identification of the most appropriate setting for a given problem and, in turn, the sensitivity test to be applied. A list of possible settings (Saltelli *et al.*, 2004) is given here:

- The **Factor Prioritization** setting (FP) is used to identify a factor (or group of factors) which, when fixed to its true value, leads to the greatest reduction in the variance of the output. In other words, the identified factor (or group of factors) is that which accounts for most of the output variance. Therefore, this setting allows us (a) to detect and rank those factors which need to be better measured in order to reduce the output variance, as well as (b) to detect the factors that have a better chance of being estimated in a subsequent numerical or experimental estimation process. This latter point is particularly interesting as the analyst can identify the factors to be estimated before any estimation is made or measurements carried out. See Tarantola *et al.* (2000) for an example in the field of physics.
- The **Factor Fixing** setting (FF) is used to identify factors in the model which, left free to vary over their range of uncertainty, make no significant contribution to the variance of the output. The identified factors can then be fixed at any given value within their range of variation without affecting the output variance. This analysis can be performed on groups of factors, especially for large models, to identify noninfluential subsets of factors. Sometimes, factors are set up to represent alternative structures for model components (e.g. simple versus complex) and significant model simplifications can be often achieved when these factors are found to be noninfluential.
- The **Variance Cutting** setting (VC) is used for the reduction of the output variance to below a given tolerance. This may be desirable in risk or reliability analysis, where the analyst is interested in making sure, for example, that the uncertainty of the reliability of a given system component is below a given tolerance. In this setting the analyst wants to guarantee that the uncertainty is brought under a given value by acting on the smallest possible number of factors (see Saltelli and Tarantola, 2002).
- The **Factor Mapping** setting (FM) is used to study which values of the input factors lead to model realizations in a given range of the output space. For example, one may want to highlight model realizations falling above the 95th percentile because these correspond to risky conditions in an industrial plant or to a considerable financial loss (Campolongo *et al.*, 2007). In this setting one investigates which combination of factors leads to the realizations under analysis (Monte Carlo filtering, see Chapter 5).

The utility of variance-based sensitivity measures derives from their wide range of application. Of the four settings just recounted, the first three are susceptible of variance-based analysis.

4.2 WHY VARIANCE?

Most models live through their operational life as 'deterministic'. Each time they are 'interrogated' they are fed with a deterministic set of values for the input variables and the output – be it a scalar, a time series or a 2D map – is investigated for possible inferences. Sensitivity analysis for these models will generally entail changing one input at a time to test its effect on the output. In this book we profess a different philosophy of modelling, in which modellers are willing – and usually eager – to explore their model over different combinations of values for the uncertain inputs. Variance-based measures have proven useful in this framework. They study how the variance of the output depends on the uncertain input factors and can be decomposed accordingly.

But why study the variance? Could a sensitivity measure be built on the mean? To give an example, in risk analysis the model output may happen to be itself a mean,[1] and we might be interested in how the mean of the model output depends on its constituents. A legitimate question would then be how much each component of a system contributes to the risk that the system might fail. The answer could be that the risk depends 25% on component A, 15% on component B, and so on. Decomposing the absolute level of the risk into system components can be useful to help understanding which component is worth improving in order to reduce the level of risk in the system.[2]

Another measure encountered in risk analysis is the 'risk reduction worth', which measures the amount by which the risk associated with a system could be reduced if a model element were perfect, i.e. without risk of failure (see Borgonovo and Apostolakis, 2001).

The two examples just illustrated are based on a deterministic output – risk – and how the risk level can be modified by eliminating the uncertainty in the input. We are in principle against such an elimination of uncertainty; we prefer rather to retain the uncertain factors as an ingredient of sensitivity analysis. Even when we study how the mean of the output changes when

---

[1] For example, a risk may happen to be estimated as the product of the probability of a given outcome and the consequence of that outcome, averaged over a set of possible outcomes.
[2] Risk analysts use, for example, the Fussell–Vesely measure of importance in probabilistic safety assessments. This measure is defined as the fraction of risk that is contributed by the failure of a model element (Borgonovo and Apostolakis, 2001).

a factor is fixed – we denoted this in Chapter 1 as $E(Y|X_i = x_i^*)$ – we would then take the variance of this measure over all possible values x_i^*, i.e. $V(E(Y|X_i))$. Taking the mean of $E(Y|X_i = x_i^*)$ would have been of scant use – it would have led us back to the overall (unconditioned) mean. In other words, in a Monte Carlo framework variance emerges naturally if one wants to preserve the factors' uncertainty.

Returning to our example of risk analysis, note that 'risk' has been expressed as a crisp figure (e.g. a rate of failure, or the expected incidence of health effects), which may distract from the fact that risk is in itself an average. A practitioner might be interested in how the average is arrived at, in the form of the risk distribution tails and in details such as the topology of the low probability high-consequence regions and in the key assumptions shaping these regions. These issues are addressed by the methods presented in this book.

A key issue in sensitivity analysis is how to quantify the uncertainty of a model prediction – variance clearly being just one of the possible options. Depending on the problem at hand, we might be interested in the model prediction falling in the upper or lower 5th percentile of the distribution or in any particular interval of interest in the distribution, as in Monte Carlo filtering.

Methods have been developed which look at the entire distribution of the output and at how this is modified, on average, if a factor is fixed (Borgonovo, 2006).

We recommend using variance as a summary measure of uncertainty whenever the application allows it. This is in order to exploit the statistical properties of variance described in this chapter to investigate how factors contribute to the variance.

Interesting features of variance-based methods are:

- model independence: the sensitivity measure is model-free;
- capacity to capture the influence of the full range of variation of each input factor;
- appreciation of interaction effects among input factors;
- capacity to tackle groups of input factors: uncertain factors might pertain to different logical types, and it might be desirable to decompose the uncertainty according to these types.

The drawback of variance-based measures is their computational cost, as we shall discuss later in the chapter, and this is the reason why much recent research aims to find efficient numerical algorithms for their computation. Techniques for computation are offered both in this chapter and the next.

Sensitivity measures based on the decomposition of the variance of the model output are relatively recent in the literature. A brief summary of their development follows.

4.3 VARIANCE-BASED METHODS. A BRIEF HISTORY

Variance-based methods for sensitivity analysis were first employed by chemists in the early 1970s (Cukier et al., 1973). Cukier and colleagues not only proposed conditional variances for a sensitivity analysis based on first-order effects, but were already aware of the need to treat higher-order terms and of the underlying variance decomposition theorems (Cukier et al., 1978). Their method, known as FAST (Fourier Amplitude Sensitivity Test), although quite effective, enjoyed limited success among practitioners, not least because of the difficulty in encoding it. The method did not allow the computation of higher-order indices, although this was much later made possible by extensions developed by other investigators (see Saltelli et al., 1999).

Also much later, Hora and Iman (1986) introduced the 'uncertainty importance' of a factor X_i,[3] defined as the expected reduction in the variance of the model output Y achieved by fixing X_i at a given value within its range of uncertainty:

$$I_i = \sqrt{\text{Var}(Y) - E[\text{Var}(Y|X_i)]}. \qquad (4.1)$$

Later, the same authors (Iman and Hora, 1990) proposed a new statistic based on estimating the following quantity:

$$\frac{\text{Var}_{X_i}[E(\log Y|X_i)]}{\text{Var}[\log Y]}. \qquad (4.2)$$

From Chapter 1 it is clear that this is the first-order variance term relative to $\log Y$. This measure has the advantage of robustness,[4] although it is not easy to convert results of sensitivity analysis pertaining to $\log Y$ back to Y. Transformations of Y for sensitivity analysis are presented in Chapter 5.

A visual inspection of sensitivity results has been suggested by Sacks et al. (1989). They proposed a decomposition of the response

$$Y = f(X_1, X_2, \ldots X_k) \qquad (4.3)$$

into a set of functions of increasing dimensionality, whose plots are themselves used as measures of sensitivity (as will be discussed in detail in

---

[3] We refer to X_i as the ith element of $\mathbf{X}$, though the formulas presented in this chapter are appropriate also if X_i corresponds to a subset of model inputs.
[4] The range of variation of $\log Y$ can be much smaller than that of Y and hence its estimate can be obtained – ceteris paribus – at a lower sample size. For the same reason formulas similar to (4.2) were proposed on the rank of Y, e.g. replacing the values of Y with the integer corresponding to 1 for the highest Y value and with N (the size of the sample) for the lowest (Saltelli et al., 1993).

Chapter 5). Although these authors do not use the variance, the functions they consider are the same as in Sobol's functional development.

The Russian mathematician I. M. Sobol' was inspired by the work of Cukier, and generalized it to provide a straightforward Monte Carlo-based implementation of the concept, capable of computing sensitivity measures for arbitrary groups of factors.

Given a square integrable function f over Ω^k, the k-dimensional unit hypercube,

$$\Omega^k = \{X | 0 \leq x_i \leq 1; i = 1, \ldots, k\}, \qquad (4.4)$$

Sobol' considers an expansion of f into terms of increasing dimensions:

$$f = f_0 + \sum_i f_i + \sum_i \sum_{j>i} f_{ij} + \cdots + f_{12\ldots k} \qquad (4.5)$$

in which each individual term is also square integrable over the domain of existence and is a function only of the factors in its index, i.e. $f_i = f_i(X_i), f_{ij} = f_{ij}(X_i, X_j)$ and so on. This decomposition is not a series decomposition, as it has a finite number of terms. Of the 2^k terms, one is constant (f_0), k are first-order functions (f_i),

$$\binom{k}{2}$$

are second-order functions (f_{ij}), and so on. This expansion, called high-dimensional model representation (HDMR), is not unique, meaning that, for a given model f, there could be infinite choices for its terms. Sobol' proved that, if each term in the expansion above has zero mean, i.e. $\int f(x_i) dx_i = 0$, then all the terms of the decomposition are orthogonal in pairs, i.e. $\int f(x_i) f(x_j) dx_i dx_j = 0$. As a consequence, these terms can be univocally calculated using the conditional expectations of the model output Y.

In particular,

$$f_0 = E(Y) \qquad (4.6)$$

$$f_i = E(Y|X_i) - E(Y) \qquad (4.7)$$

$$f_{ij} = E(Y|X_i, X_j) - f_i - f_j - E(Y) \qquad (4.8)$$

An analytic example of this decomposition is offered in Exercise 5.

As mentioned in Chapter 1 (see Conditional Variances – second path), the conditional expectation $E(Y|X_i)$ can be calculated empirically by cutting the X_i domain into slices and averaging the values of $(Y|X_i)$ within the same slice X_i. In this way, if the scatterplot has a pattern, the conditional expectation $E(Y|X_i)$ has a large variation across X_i values and the factor X_i is revealed to be important. Hence, the variance of the conditional expectation can be considered as a summary measure of sensitivity. In fact, the variances of the terms in the decomposition above are the measures of importance being sought. In particular, $V(f_i(X_i))$ is $V[E(Y|X_i)]$; when we divide this by the unconditional variance $V(Y)$, we obtain the first-order sensitivity index. In short:

$$S_i = \frac{V[E(Y|X_i)]}{V(Y)}. \tag{4.9}$$

The first-order index represents the main effect contribution of each input factor to the variance of the output. The same quantity has been described by different investigators as an 'importance measure' (see Hora and Iman, 1986; Ishigami and Homma, 1996; Iman and Hora, 1990; Saltelli et al., 1993; Homma and Saltelli, 1996), and as a 'correlation ratio' (see Krzykacz-Hausmann, 1990; McKay, 1996).

Sobol' also proposed a comparable definition of S_i (Sobol', 1996) which is based on the correlation between the model Y and the conditional expectation $E(Y|X_i)$:

$$S_i = Corr(Y, E(Y|X_i)). \tag{4.10}$$

4.4 INTERACTION EFFECTS

How can Sobol's variance decomposition help in investigating the existence of interaction effects? Two factors are said to interact when their effect on Y cannot be expressed as a sum of their single effects. Interactions may imply, for instance, that extreme values of the output Y are uniquely associated with particular combinations of model inputs, in a way that is not described by the first-order effects S_i just mentioned. Interactions represent important features of models, and are more difficult to detect than first-order effects. For example, by using regression analysis tools it is fairly easy to estimate first-order indices, but not interactions (remember the relationship $S_i = \beta_{X_i}^2$ discussed in Chapter 1 for linear models and orthogonal inputs, where β_{X_i} is the standardized regression coefficient for factor X_i).

A useful feature of decomposition (4.7 and 4.8) is that

$$V_i = V(f_i(X_i)) = V[E(Y|X_i)]$$

and

$$V_{ij} = V(f_{ij}(X_i, X_j)) = V(E(Y|X_i, X_j)) - V(E(Y|X_i)) - V(E(Y|X_j)).$$

In this equation, $V(E(Y|X_i, X_j))$ measures the joint effect of the pair (X_i, X_j) on Y, and, from now on we will denote the joint effect by V_{ij}^c. The term

$V(f_{ij})$ is the joint effect of X_i and X_j minus the first-order effects for the same factors. $V(f_{ij})$ is known as a second-order, or two-way, effect (Box et al., 1978). Analogous formulas can be written for higher-order terms, enabling the analyst to quantify the higher-order interactions.

By condensing the notation of the variances, i.e. $V(f_i) = V_i$, $V(f_{ij}) = V_{ij}$, and so on, and by square integrating each term of the decomposition (4.5) over Ω^k, we can write the so-called ANOVA-HDMR decomposition:[5]

$$V(Y) = \sum_i V_i + \sum_i \sum_{j>i} V_{ij} + \ldots + V_{12\ldots k}. \qquad (4.11)$$

Dividing both sides of the equation by $V(Y)$, we obtain

$$\sum_i S_i + \sum_i \sum_{j>i} S_{ij} + \sum_i \sum_{j>i} \sum_{l>j} S_{ijl} + \ldots + S_{123\ldots k} = 1 \qquad (4.12)$$

which we already know from Chapter 1.

We recall that the number of these terms increases exponentially with the number of input factors.

Exercise 5, part 2, provides an example of the computation of partial variances and sensitivity indices for the same analytic case used for the functional decomposition in Section 4.3.

4.5 TOTAL EFFECTS

Total effects are a direct consequence of Sobol's variance decomposition approach and estimation procedure, although they were explicitly introduced and made computationally affordable by other investigators (see Homma and Saltelli, 1996; Saltelli, 2002).

The total effect index accounts for the total contribution to the output variation due to factor X_i, i.e its first-order effect plus all higher-order effects due to interactions.

For a three-factor model, for example, the total effect of X_1 is the sum of all the terms in Equation (4.12) where the factor X_1 is considered:

$$S_{T1} = S_1 + S_{12} + S_{13} + S_{123}. \qquad (4.13)$$

---

[5] This is because of the orthogonality properties between any pair of terms in the expansion (see Exercise 5, part 3). Note that this variance decomposition holds only when the input factors X_i are independent (i.e. orthogonal). When the input factors are not independent of one another, the quantities V_i, V_{ij}^c, V_{ijl}^c retain their meaning but are no longer related to one another via (4.11).

TOTAL EFFECTS

In this example, the total index is composed of four terms. Total indices are useful in sensitivity analysis, as they give information on the nonadditive features of the model. As mentioned, for a purely additive model $\sum_{i=1}^{k} S_i = 1$, while for a given factor X_i a significant difference between S_{Ti} and S_i signals important interaction involving that factor. The total indices could be calculated in principle by computing all the terms in the decomposition (4.12), but there are as many as $2^k - 1$ of these. There are techniques that enable us to estimate total indices at the same cost of first-order indices (such as the Sobol' technique, see Homma and Saltelli (1996)), thus circumventing the so-called 'curse of dimensionality'. We customarily compute the set of all S_i plus the set of all S_{Ti} to obtain a fairly good description of the model sensitivities at a reasonable cost. We will see how to compute these indices in the next section.

The total effect measure provides the educated answer to the question: 'Which factor can be fixed anywhere over its range of variability without affecting the output?' The condition $S_{Ti} = 0$ is necessary and sufficient for X_i to be a noninfluential factor. If $S_{Ti} \cong 0$, then X_i can be fixed at any value within its range of uncertainty without appreciably affecting the value of the output variance $V(Y)$. The approximation error that is made when this model simplification is carried out depends on the value of S_{Ti} (see Sobol' et al., 2007). Total indices are suitable for the factor fixing setting.

We recall from Chapter 1 that the unconditional variance can be decomposed into main effect and residual:

$$V(Y) = V(E(Y|X_i)) + E(V(Y|X_i)). \quad (4.14)$$

Another way to find the total index is to decompose the output variance $V(Y)$ again, in terms of main effect and residual, conditioning this time with respect to all the factors but one, i.e. $\mathbf{X}_{\sim i}$:

$$V(Y) = V(E(Y|\mathbf{X}_{\sim i})) + E(V(Y|\mathbf{X}_{\sim i})). \quad (4.15)$$

The measure $V(Y) - V(E(Y|\mathbf{X}_{\sim i})) = E(V(Y|\mathbf{X}_{\sim i}))$ is the remaining variance of Y that would be left, on average, if we could determine the true values of $\mathbf{X}_{\sim i}$. The average is calculated over all possible combinations of $\mathbf{X}_{\sim i}$, since $\mathbf{X}_{\sim i}$ are uncertain factors and their 'true values' are unknown. Dividing by $V(Y)$ we obtain the total effect index for X_i:

$$S_{T_i} = \frac{E[V(Y|\mathbf{X}_{\sim i})]}{V(Y)} = 1 - \frac{V[E(Y|\mathbf{X}_{\sim i})]}{V(Y)}. \quad (4.16)$$

4.6 HOW TO COMPUTE THE SENSITIVITY INDICES

In this section we describe the Monte-Carlo based numerical procedure for computing the full set of first-order and total-effect indices for a model of k factors.

This procedure is the best available today for computing indices based purely on model evaluations. Additional procedures are described in the next chapter, based on emulators, i.e. on the ability to generate estimates of model output at untried points without rerunning the simulation model. The method offered here is attributable to Saltelli (2002) and represents an extension of the original approach provided by Sobol' (1990) and Homma and Saltelli (1996).

At first sight, it might seem that the computational strategy for the estimation of conditional variances such as $V(E(Y|X_i))$ and $V(E(Y|X_i, X_j))$ would be the cumbersome, brute-force computation of the multidimensional integrals in the space of the input factors. To obtain, for example, $V(E(Y|X_i))$, one would first use a set of Monte Carlo points to estimate the inner expectation for a fixed value of X_i, and then repeat the procedure many times for different X_i values to estimate the outer variance. To give an indication, if 1000 points were used to get a good estimate of the conditional mean $E(Y|X_i)$, and the procedure were repeated 1000 times to estimate the variance, then we would need 10^6 points just for one sensitivity index.

This is in fact not necessary, as the computation can be accelerated via existing short cuts. In the following, we describe the instrument proposed by Saltelli:

- Generate a (N, 2k) matrix of random numbers (k is the number of inputs) and define two matrices of data (A and B), each containing half of the sample (see 4.17 and 4.18). N is called a base sample; to give an order of magnitude, N can vary from a few hundreds to a few thousands. Sobol' recommends using sequences of quasi-random numbers (Sobol', 1967, 1976). The software to generate these sequences is freely available (**SIMLAB**, 2007).

$$A = \begin{bmatrix} x_1^{(1)} & x_2^{(1)} & \cdots & x_i^{(1)} & \cdots & x_k^{(1)} \\ x_1^{(2)} & x_2^{(2)} & \cdots & x_i^{(2)} & \cdots & x_k^{(2)} \\ \cdots & \cdots & \cdots & \cdots & & \\ x_1^{(N-1)} & x_2^{(N-1)} & \cdots & x_i^{(N-1)} & \cdots & x_k^{(N-1)} \\ x_1^{(N)} & x_2^{(N)} & \cdots & x_i^{(N)} & \cdots & x_k^{(N)} \end{bmatrix} \quad (4.17)$$

HOW TO COMPUTE THE SENSITIVITY INDICES

$$B = \begin{bmatrix} x_{k+1}^{(1)} & x_{k+2}^{(1)} & \cdots & x_{k+i}^{(1)} & \cdots & x_{2k}^{(1)} \\ x_{k+1}^{(2)} & x_{k+2}^{(2)} & \cdots & x_{k+i}^{(2)} & \cdots & x_{2k}^{(2)} \\ \cdots & \cdots & \cdots & \cdots & \cdots & \cdots \\ x_{k+1}^{(N-1)} & x_{k+2}^{(N-1)} & \cdots & x_{k+i}^{(N-1)} & \cdots & x_{2k}^{(N-1)} \\ x_{k+1}^{(N)} & x_{k+2}^{(N)} & \cdots & x_{k+i}^{(N)} & \cdots & x_{2k}^{(N)} \end{bmatrix}. \qquad (4.18)$$

- Define a matrix C_i formed by all columns of B except the ith column, which is taken from A:

$$C_i = \begin{bmatrix} x_{k+1}^{(1)} & x_{k+2}^{(1)} & \cdots & x_i^{(1)} & \cdots & x_{2k}^{(1)} \\ x_{k+1}^{(2)} & x_{k+2}^{(2)} & \cdots & x_i^{(2)} & \cdots & x_{2k}^{(2)} \\ \cdots & \cdots & \cdots & \cdots & \cdots & \cdots \\ x_{k+1}^{(N-1)} & x_{k+2}^{(N-1)} & \cdots & x_i^{(N-1)} & \cdots & x_{2k}^{(N-1)} \\ x_{k+1}^{(N)} & x_{k+2}^{(N)} & \cdots & x_i^{(N)} & \cdots & x_{2k}^{(N)} \end{bmatrix}. \qquad (4.19)$$

- Compute the model output for all the input values in the sample matrices A, B, and C_i, obtaining three vectors of model outputs of dimension $N \times 1$:

$$y_A = f(A) \qquad y_B = f(B) \qquad y_{C_i} = f(C_i). \qquad (4.20)$$

We anticipate that these vectors are all we need to compute the first- and total-effect indices S_i and S_{T_i}, for a given factor X_i. Because there are k factors, the cost of this approach is $N+N$ runs of the model for matrices A, B, plus k times N to estimate k times the output vector corresponding to matrix C_i. The total cost is hence $N(k+2)$, much lower than the N^2 runs of the brute-force method.

Our recommended method estimates first-order sensitivity indices as follows:

$$S_i = \frac{V[E(Y|X_i)]}{V(Y)} = \frac{y_A \cdot y_{C_i} - f_0^2}{y_A \cdot y_A - f_0^2} = \frac{(1/N)\sum_{j=1}^N y_A^{(j)} y_{C_i}^{(j)} - f_0^2}{(1/N)\sum_{j=1}^N (y_A^{(j)})^2 - f_0^2} \qquad (4.21)$$

where

$$f_0^2 = \left(\frac{1}{N}\sum_{j=1}^N y_A^{(j)}\right)^2 \qquad (4.22)$$

is the mean, and the symbol $(\cdot)$ denotes the scalar product of two vectors.
Similarly, the method estimates total-effect indices as follows:

$$S_{T_i} = 1 - \frac{V[E(Y|\mathbf{X}_{\sim i})]}{V(Y)} = 1 - \frac{y_B \cdot y_{C_i} - f_0^2}{y_A \cdot y_A - f_0^2} = 1 - \frac{(1/N)\sum_{j=1}^N y_B^{(j)} y_{C_i}^{(j)} - f_0^2}{(1/N)\sum_{j=1}^N (y_A^{(j)})^2 - f_0^2}. \qquad (4.23)$$

Why do these formulas work? We offer here a 'hand waving' explanation of (4.21). In the scalar product $y_A \cdot y_{C_i}$ values of Y computed from A are multiplied by values of Y for which all factors but X_i are resampled while the values of X_i remain fixed. If X_i is noninfluential, then high and low values of y_A and y_{C_i} are randomly associated. If X_i is influential, then high (or low) values of y_A will be preferentially multiplied by high (or low) values of y_{C_i} increasing the value of the resulting scalar product. We leave to the reader the task of understanding (4.23). (Hint: the scalar product $y_B \cdot y_{C_i}$ gives the first-order effect of non-X_i.)

Note that the accuracy of both f_0 and $V(Y)$ can be improved by using both y_A and y_B points rather than just y_A in Equations (4.21) and (4.23) (see Saltelli, 2002). This will improve the accuracy of the estimates for S_i and S_{T_i}, although the factors' ranking will remain unchanged. Error estimates for Equations (4.21) and (4.23) can be obtained by bootstrapping data points from vectors y_A, y_B and y_{C_i}. Alternatively, the error in the numerical estimates can be evaluated using the probable error associated with the crude Monte Carlo estimate.

The probable error is the error which will not be exceeded by 50 percent of the cases, and corresponds to 0.6745 [6] times the standard error.

For example, the probable error in $V_{X_i}[E(Y|X_i)]$ (that will not be exceeded by the error in the estimate with 50% probability) is:

$$P.E. = \frac{0.6745}{\sqrt{N}} \sqrt{\sum_{j=1}^{N}(y_A^{(j)}y_{C_i}^{(j)})^2 - \left(\sum_{j=1}^{N}y_A^{(j)}y_{C_i}^{(j)}\right)^2}.$$

Before using the sensitivity measures in our case studies, let us recall some of the properties of sensitivity indices that will prove useful in the interpretation of the results:

- Whatever the strength of the interactions in the model, S_i indicates by how much one could reduce, on average, the output variance if X_i could be fixed; hence, it is a measure of main effect.
- Whatever the interactions in the model, $S^c_{i_1,i_2,\ldots,i_s}$ indicates by how much the variance could be reduced, on average, if one could fix $X_{i_1}, X_{i_2}, \ldots, X_{i_s}$. We recall that 'c' denotes the joint effect.
- By definition, S_{T_i} is greater than S_i, or equal to S_i in the case that X_i is not involved in any interaction with other input factors. The difference

---

[6] The probability P_δ that a random sample from a normally distributed universe will have a mean m within a distance $|\delta|$ of the mean μ of the universe is $P_\delta = 2\Phi(|\delta|)$ where $\Phi(z)$ is the standard normal distribution function and δ is the observed value of $z = \frac{\overline{X}-\mu}{\frac{\sigma}{\sqrt{N}}}$.
The value δ^* of δ such that $P_\delta = \frac{1}{2}$, is given by $\delta^* = \sqrt{2} \cdot erf^{-1}\left(\frac{1}{2}\right) = 0.6745$.

$S_{T_i} - S_i$ is a measure of how much X_i is involved in interactions with any other input factor.
- $S_{T_i} = 0$ implies that X_i is noninfluential and can be fixed anywhere in its distribution without affecting the variance of the output.
- The sum of all S_i is equal to 1 for additive models and less than 1 for nonadditive models. The difference $1 - \sum_i S_i$ is an indicator of the presence of interactions in the model.
- The sum of all $S_{T_i}s$ is always greater than 1. It is equal to 1 if the model is perfectly additive.

4.7 FAST AND RANDOM BALANCE DESIGNS

The classic Fourier Amplitude Sensitivity Test (FAST) method (Cukier *et al.*, 1978) is based on selecting N design points over a particular space-filling curve in the kth dimensional input space, built so as to explore each factor with a different (integer) frequency ($\omega_1, \omega_2, \ldots, \omega_k$). A quite complex algorithm is used to set the frequencies such that they are free of interferences up to a given order M ($M = 6$ is usually considered sufficient). The computational model is run at each design point and the Fourier spectrum is calculated on the model output at specific frequencies ($\omega_i, 2\omega_i, \ldots M\omega_i$) to estimate the sensitivity index of factor X_i. It is important that none of the higher harmonics of $\omega_1, \omega_2, \ldots, \omega_k$ interfere until order M, so that the Fourier spectrum at a given frequency corresponds uniquely to factor X_i. The design points are selected as follows:

$$X_i(s_j) = G_i(\sin \omega_i s_j), \quad \forall i = 1, 2, ..k, \quad \forall j = 1, 2, .., N \quad (4.24)$$

where X_i is the ith input factor, the functions G_i are chosen according to the desired pdf of X_i, s_j is the parametric variable varying in $(-\pi, \pi)$ which is sampled over its range using N points, and ω_i are the frequencies.

In Random Balance Designs (RBD) (Tarantola *et al.*, 2006) N points are selected over a curve in the input space using a frequency equal to 1 for each factor. The curve covers only a subset of the input space. Then independent random permutations are applied to the coordinates of the N points in order to generate the design points. The computational model is evaluated at each design point. Subsequently, the model outputs are reordered such that the design points are in increasing order with respect to factor X_i. The Fourier spectrum is calculated on the model output at the frequency 1 and at its higher harmonics (2, 3, 4, 5, 6) and yields the estimate of the sensitivity index of factor X_i. The same model outputs are reordered with respect to each other factor (and the Fourier spectra are calculated accordingly) to obtain all the other sensitivity indices. The design points are chosen as follows:

$$X_i(s_{ij}) = G_i(\sin \omega s_{ij}), \quad \forall i = 1, 2, ..k, \quad \forall j = 1, 2, .., N \quad (4.25)$$

where $(s_{i1}, s_{i2}, \ldots, s_{iN})$ denotes the ith random permutation of the N points. Equation (4.25) provides a different random permutation for each factor X_i.

For RBD the model is evaluated N times over the sample of size N:

$$Y(s_j) = f(X_1(s_{1j}), X_2(s_{2j}), \ldots, X_k(s_{kj})) \quad \forall j = 1, 2, .., N. \quad (4.26)$$

The values of model output $Y(s_j)$, j = 1,..N are then reordered ($Y^R(s_j)$) such that the corresponding values of $X_i(s_{ij})$ are ranked in increasing order. The sensitivity of Y to X_i is quantified by the Fourier spectrum of the reordered model output:

$$F(\omega) = \left| \frac{1}{\pi} \sum_{j=1}^{N} Y^R(s_j) \exp(-i\omega s_j) \right|^2 \quad (4.27)$$

evaluated at $\omega = 1$ and its higher harmonics (2, 3, 4, 5, 6). In the discrete case:

$$\widehat{V}_i = V[E(Y|X_i)] = \sum_{l=1}^{M} F(\omega)|_{\omega=l} = \sum_{l=1}^{M} F(l). \quad (4.28)$$

This is an estimate of the main effect V_i. The procedure is repeated for all factors, whereby the same set of model outputs is simply reordered according to $X_i(s_{ij})$ and (4.27) and (4.28) are used to estimate V_i, i= 2,..., k.

Here we provide the basic *Matlab*® code to compute a generic sensitivity index according to the RBD method:

```
s0=[-pi:2*pi/N:pi]';
s=s0(randperm(N))'; Performs a random permutation of the integers from 1 to N
x=.5+asin(sin(1*s))/pi; (see (4.25))
[dummy,index]=sort(s); orders the elements of s in ascending order
Y=model(x)
yr=y(index);
spectrum=(abs(fft(yr))).^2/N; fft is the fast Fourier transform
V1=2*sum(spectrum(2:M+1)); (see 4.28)
V=sum(spectrum(2:N));
S1=V1/V;
```

Random Balance Designs have a number of advantages with respect to FAST:

THE INFECTION DYNAMICS MODEL

- the absence of a lower limit for the size N of the design points (FAST has the problem of aliasing, so a minimum sample size is required and this minimum size increases with the dimensionality k);
- the nonnecessity to have an algorithm to search for frequencies free of interferences;
- better accuracy in the estimates, which are not influenced by interferences;
- the possibility to select larger values of the order M without affecting the sample size N.
- contrary to the method of Saltelli, each model run contributes to the estimation of all the first-order indices.

The disadvantage of the RBD method is that it allows the computation of first-order terms only; we can use the sum of these to check if the model is additive. If the sum is noticeably smaller than 1 we must use another algorithm to compute interactions or total-effect terms. The main advantage of RBD is that it is relatively easy to implement, and the sample size N, being independent of the number of factors k, can lead to a considerable saving in computer time for expensive models.

4.8 PUTTING THE METHOD TO WORK: THE INFECTION DYNAMICS MODEL

Let us consider an infective process at its early stage, where I is the number of infected individuals at time t and S is the number of individuals susceptible to infection at time t.

We assume that the infection is propagated through some kind of contact between individuals who, especially at the early stage, do not take any precaution to avoid contagion.

It is reasonable to assume that the number of contacts per unit time is proportional to the number of individuals in each group (i.e. to $I \times S$) via a contact coefficient $k < 1$. Also, the number of infections is proportional to the number of contacts through an 'infection coefficient' ($\gamma < 1$), which is the likelihood that the infection is passed on during a given contact.

Depending on the dangerousness of the infection, the infected individuals will end up in either of two ways: by recovering or by dying. It is presumed that recovery and death rates (r and δ) are proportional to the number of infected individuals.

The number of susceptible individuals decreases with the number of infections, but can increase with new births b, proportional to S, or migration which happens at a constant rate m.

Two equations describe the dynamics of I and S, representing the model of the infection process:

$$\frac{dI}{dt} = \gamma k I S - rI - \delta I \quad (4.29)$$

$$\frac{dS}{dt} = -\gamma k I S + bS + m. \quad (4.30)$$

Let us investigate the evolution of the infection at its early stage $t \sim 0$, when we presuppose that the number of the susceptible individuals is much larger than that of the infected ($S(t) >> I(t)$), and that S is changing slowly ($S(t) \sim S_0 = const$).

Equation (4.29) becomes linear and homogeneous:

$$\frac{dI}{dt} = (\gamma k S_0 - r - \delta)I. \quad (4.31)$$

The solution is $I = I_0 \cdot exp(Y)$, where $Y = \gamma k S_0 - r - \delta$. If $Y > 0$ the infection spreads, while if $Y < 0$ the infection dies out.

Suppose that $S_0 = 1000$ (a small village), and that factors are distributed as follows:

- Infection coefficient $\gamma \sim U(0, 1)$. The infection is at an early stage, and no information is available about how it is acting.
- Contact coefficient $k \sim beta(2, 7)$. This distribution describes the probability of a person to come into contact with other individuals. In other words, the probability of meeting all the inhabitants of the village (and of meeting nobody) is low, while the probability of meeting an average number of persons is higher.
- Recovery rate $r \sim U(0, 1)$. We assume this to be uniform, as we do not know how it behaves at the beginning of the propagation.
- Death rate $\delta \sim U(0, 1)$, for the same reason as r.

Let us calculate the sensitivity indices for the four factors using the method described in Section 4.6. The total number of model runs is $N(k+2)=7680$ ($N = 1280$ and $k = 4$).

Let us analyse the sensitivity indices shown in Table 4.1. Negative signs are due to numerical errors in the estimates. Such negative values can often be encountered for the Saltelli method when the analytical sensitivity indices are close to zero (i.e. for unimportant factors). Increasing the sample size of the analysis reduces the probability of having negative estimates. FAST and RBD estimates are always positive, by construction.

The most influential factors are γ and k, while factors r and δ are noninfluential as their total indices are negligible. The infection is likely to

THE INFECTION DYNAMICS MODEL

Table 4.1 First-order and total-effect sensitivity indices obtained with the method of Saltelli with $k \sim \text{beta}(2, 7)$

Factor	First-order indices	Total-order indices
Infection (γ)	0.49	0.69
Contact (k)	0.41	0.61
Recovery (r)	−0.00	−0.00
Death (δ)	−0.00	−0.00

spread proportionally to the number of contacts between people; unlike γ, which depends on the virus strength, k is a controllable factor. This means that if the number of contacts could be reduced, the variability in the output would also be reduced.

The sum of first-order effects is approximately 0.89, while the sum of the total indices is 1.29; as these two sums are both different from 1, there must be interactions among factors in the model. Moreover, given that both factors γ and k have total indices greater than their first orders, we conclude that they are taking part in interactions.

With the model outputs calculated at the 1280 points sampled above, we perform uncertainty analysis (see Figure 4.1, case 1). The picture shows that the infection propagates in almost all cases (in 99.7% of the cases the sign of the model predictions is positive).

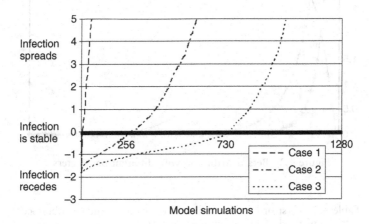

Figure 4.1 Uncertainty analysis for the three cases of the infection dynamics exercise. On the Y axis we plot the output variable of interest (if Y is positive the infection propagates, if it is negative the infection recedes and if it is zero – the X axis – the number of infected individuals is stable), and on the X axis the total number of model runs. Model outputs for each case are sorted in increasing order, so that each plot is a monotonic curve. As almost all model runs for the first case correspond to positive Y's, many of which can be of the order of a hundred, the Y axis is cut at +5 to visualize the plot around zero

Assume that, in consequence of this epidemic spread, some measures are taken in order to reduce the propagation of the disease. Following these new measures, which warn the inhabitants to avoid contacts, we assume that k is now distributed as $k \sim beta(0.5, 10)$ (see Figure 4.2) and we repeat the sensitivity analysis to see how the relative importance of the factors has been modified. The results are reported in Table 4.2.

We observe that, in this new configuration, factor k becomes more important in controlling the spread of the infection.

This causes us to wonder whether the restriction is adequate to reduce the propagation of the infection. By looking at the output of the model, we see that the infection recedes in 20% of cases, while in 80% of cases it propagates (see Figure 4.1, case 2): that is a significant improvement with respect to the initial situation, but stronger measures could still be taken.

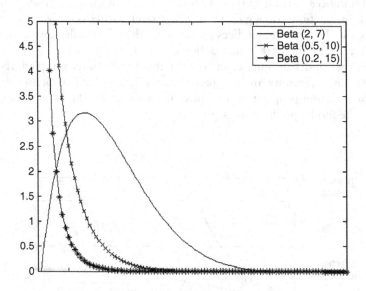

Figure 4.2 Beta distribution with different parameters

Table 4.2 First-order and total effect sensitivity indices obtained with the method of Saltelli with $k \sim \text{beta}(0.5, 10)$

Factor	First-order indices	Total-order indices
Infection (γ)	0.14	0.36
Contact (k)	0.76	0.98
Recovery (r)	−0.00	−0.00
Death (δ)	−0.00	−0.00

THE INFECTION DYNAMICS MODEL

Assume that we implement additional restrictions on contact between persons, which means further squeezing the distribution of k to the left-hand side of its uncertainty range ($k \sim beta(0.2, 15)$, see Figure 4.2).

We observe in Table 4.3 that factor k becomes even more important, while factor γ has less influence in controlling the spread of the infection. The uncertainty analysis shows that in 57% of cases the infection recedes (see Figure 4.1, case 3).

The present example is quite academic, yet it shows how information obtained from sensitivity analysis (e.g. that the amount of contact between people is the most important factor in determining the spread of the disease) can help to inform decisions (e.g. designing measures to reduce people's contact in order to control the infection's propagation).

We test the RBD method described in Section 4.7 on the same case study selecting $N = 1280$, i.e. the same sample size that was employed for the method of Saltelli. In the RBD method each model run contributes to the estimation of all the sensitivity indices, while in the method of Saltelli it contributes to the estimation of one single first-order index (and its related total effect). In summary, RBD has better convergency properties than the method of Saltelli, in the sense that, for a given sample size, RBD estimates are more accurate (Tarantola et al., 2006). We report the results for the three configurations of k in Table 4.4.

RBD produces indices for important factors (i.e. infection and contact) which are similar to those obtained with the method of Saltelli. Although the nonrelevant factors (recovery and death) are somewhat overestimated

Table 4.3 First-order and total-effect sensitivity indices obtained with the method of Saltelli with $k \sim beta(0.2, 15)$

Factor	First-order indices	Total-order indices
Infection (γ)	0.05	0.32
Contact (k)	0.77	1.05
Recovery (r)	−0.00	−0.00
Death (δ)	0.00	0.00

Table 4.4 RBD method first-order sensitivity indices

Factor	$k \sim beta(2, 7)$	$k \sim beta(0.5, 10)$	$k \sim beta(0.2, 15)$
Infection (γ)	0.43	0.13	0.05
Contact (k)	0.41	0.58	0.65
Recovery (r)	0.01	0.01	0.01
Death (δ)	0.01	0.01	0.01

by RBD, this is a minor problem, since they can anyhow be identified as noninfluential.

4.9 CAVEATS

Variance-based methods are powerful in quantifying the relative importance of input factors or groups. The main drawback of variance-based methods is the cost of the analysis, which, in the case of computationally intensive models, can become prohibitive even when using the approach described above.

With Saltelli's method, $N(k+2)$ runs for a full set of S_i and S_{Ti} require that for a model with 15 factors we need to execute the model at least 17 000 times, taking $N = 1000$. Using random balance designs with just N model executions we can compute the full set of S_i, but the S_{Ti} would remain unknown.

In terms of computational time, thousands or tens of thousands of model executions can be either trivial or unfeasible, depending on the model at hand. A viable alternative for computationally expensive models is the screening method discussed in Chapter 3. The elementary effect test is a good proxy for the total sensitivity indices.

If the model is both expensive to run and rich in factors we recommend using the elementary effect method to reduce the number of factors and then running a variance-based analysis on a reduced set of factors.

4.10 EXERCISES

Exercise 1

Let us consider the model $Y = \sum_{j=1}^{k} X_j$ where $k=3$, $X_j \sim U(\overline{x_j} - \sigma_j, \overline{x_j} + \sigma_j)$, $\overline{x_j} = 3^{j-1}$ and $\sigma_j = 0.5\overline{x_j}$.

Calculate the first-order sensitivity indices for the k factors.

First we calculate the expected value and the variance for the model.

$$Y = X_1 + X_2 + X_3$$
$$\overline{X_1} = 1$$
$$\overline{X_2} = 3$$
$$\overline{X_3} = 9$$
$$\sigma_1 = 0.5$$
$$\sigma_2 = 1.5$$

EXERCISES

$$\sigma_3 = 4.5$$
$$X_1 \sim U(0.5, 1.5)$$
$$X_2 \sim U(1.5, 4.5)$$
$$X_3 \sim U(4.5, 13.5)$$
$$E(Y) = E(X_1) + E(X_2) + E(X_3) = 1 + 3 + 9 = 13.$$

Next we compute the variance as

$$V(X_i) = E(X_i)^2 - E^2(X_i) = p(X_i) \int_a^b X_i^2 dx_i - E^2(X_i). \qquad (4.32)$$

So in our case we will have

$$V(X_1) = \int_{0.5}^{1.5} X_1^2 dx_1 - 1 = \frac{1}{12}$$

$$V(X_2) = \frac{1}{3} \int_{1.5}^{4.5} X_2^2 dx_2 - 9 = \frac{3}{4}$$

$$V(X_3) = \frac{1}{9} \int_{4.5}^{13.5} X_3^2 dx_3 - 81 = \frac{27}{4}$$

$$V(Y) = V(X_1) + V(X_2) + V(X_3) = \frac{1}{12} + \frac{3}{4} + \frac{27}{4} = \frac{91}{12}.$$

We compute the variance of the conditional expectation $V_{X_j}[E(Y|X_j)]$ and the expected residual variance $E_{X_j}[V(Y|X_j)]$.

$$V_{X_1}[E(Y|X_1)] = V(X_1) = \frac{1}{12}$$

$$V_{X_2}[E(Y|X_2)] = V(X_2) = \frac{9}{12}$$

$$V_{X_3}[E(Y|X_3)] = V(X_3) = \frac{81}{12}$$

$$E_{X_1}[V(Y|X_1)] = V(Y) - V_{X_1}[E(Y|X_1)] = \frac{91}{12} - \frac{1}{12} = \frac{90}{12}$$

$$E_{X_2}[V(Y|X_2)] = V(Y) - V_{X_2}[E(Y|X_2)] = \frac{91}{12} - \frac{9}{12} = \frac{82}{12}$$

$$E_{X_3}[V(Y|X_3)] = V(Y) - V_{X_3}[E(Y|X_3)] = \frac{91}{12} - \frac{81}{12} = \frac{10}{12}.$$

Now we have all what we need to find the first-order indices:

$$S_1 = \frac{V_{X_1}[E(Y|X_1)]}{V(Y)} = \frac{1/12}{91/12} = \frac{1}{91}$$

$$S_2 = \frac{V_{X_2}[E(Y|X_2)]}{V(Y)} = \frac{9/12}{91/12} = \frac{9}{91}$$

$$S_3 = \frac{V_{X_3}[E(Y|X_3)]}{V(Y)} = \frac{81/12}{91/12} = \frac{81}{91}.$$

The model is additive, which means that there are no interactions among factors.

Exercise 2

Consider now the model $Y = X_1 + X_2$ where X_1, X_2 are normally distributed. We also know that $\overline{x_1} = 1, \overline{x_2} = 2$ and $\sigma_1 = 2, \sigma_2 = 3$.

Compute the expected value and the variance for model Y.

$$E(Y) = E(X_1) + E(X_2) = 1 + 2 = 3.$$

In this case, as factors are normally distributed, we can calculate the variance in an easier way:

$$V(Y) = V(X_1) + V(X_2) = \sigma_1^2 + \sigma_2^2 = 4 + 9 = 13. \quad (4.33)$$

Calculate the first-order sensitivity indices for the two factors:

$$S_1 = \frac{V_{X_1}[E(Y|X_1)]}{V(Y)} = \frac{4}{13}$$

$$S_2 = \frac{V_{X_2}[E(Y|X_2)]}{V(Y)} = \frac{9}{13}.$$

Also in this case the model is additive, without interactions among factors.

Exercise 3

Two input factors are normally distributed in the model

$$Y = X_1 \times X_2.$$

with parameters $\mu_1 = 1$, $\mu_2 = 2$, $\sigma_1 = 3$ and $\sigma_2 = 2$.

Calculate the first- and second-order indices for the inputs and comment on the level of additivity of the model.

$$V(Y) = \mu_{X_1}^2 \sigma_{X_2}^2 + \mu_{X_2}^2 \sigma_{X_1}^2 + \sigma_{X_1}^2 \sigma_{X_2}^2 = 76.$$

$$S_{X_1} = \frac{\mu_{X_2}^2 \sigma_{X_1}^2}{V(Y)} = \frac{9}{19}$$

EXERCISES

$$S_{X_2} = \frac{\mu_{X_1}^2 \sigma_{X_2}^2}{V(Y)} = \frac{1}{19}$$

$$S_{X_1,X_2} = \frac{\sigma_{X_1}^2 \sigma_{X_2}^2}{V(Y)} = \frac{9}{19}.$$

Factor X_1 is the most influential in determining the output variance (its first-order index is high). Factor X_2 has a low first-order index and is thus apparently less important.

Yet the interaction effect S_{X_1,X_2} is as high as the first-order effect of factor X_1. This means that the output variance is significantly driven by the two factors' interaction, even if factor X_2 appears to be noninfluential. This shows that ignoring interactions could lead to serious type II errors.[7]

Exercise 4

A model has eight input factors, but for computational cost's reasons we need to reduce the number of factors to five.

The model is

$$Y = \sum_{i=1}^{8} X_i$$

where X_i are normally distributed as follows:

$$X_1 \sim N(0, 1)$$
$$X_2 \sim N(2, 2)$$
$$X_3 \sim N(1, 3)$$
$$X_4 \sim N(1, 5)$$
$$X_5 \sim N(3, 1)$$
$$X_6 \sim N(4, 1)$$
$$X_7 \sim N(1, 2)$$
$$X_8 \sim N(5, 5)$$

1. Calculate the first-order sensitivity indices and identify the three least important factors, in order to exclude them from the model.
2. Recalculate the first orders for the remaining five factors and find out which are the most influential: if we decide to fix them at a given value in their range of variation, by what amount will the variance of the output be reduced?

---

[7] That is, ignoring the influence of an influential factor. This is typically the most serious, nonconservative error. On the other hand type I error means considering a noninfluential factor as influential.

VARIANCE-BASED METHODS

1. We first calculate the output variance of the model:

$$V(Y) = \sum_{i=1}^{8} V(X_i) = 1+4+9+25+1+1+4+25 = 70.$$

Now it is easy to derive the first-order sensitivity indices for all input factors:

$$S_1 = \frac{V_{X_1}[E(Y|X_1)]}{V(Y)} = \frac{1}{70} = 0.01$$

$$S_2 = \frac{V_{X_2}[E(Y|X_2)]}{V(Y)} = \frac{4}{70} = 0.06$$

$$S_3 = \frac{V_{X_3}[E(Y|X_3)]}{V(Y)} = \frac{9}{70} = 0.13$$

$$S_4 = \frac{V_{X_4}[E(Y|X_4)]}{V(Y)} = \frac{25}{70} = 0.36$$

$$S_5 = \frac{V_{X_5}[E(Y|X_5)]}{V(Y)} = \frac{1}{70} = 0.01$$

$$S_6 = \frac{V_{X_6}[E(Y|X_6)]}{V(Y)} = \frac{1}{70} = 0.01$$

$$S_7 = \frac{V_{X_7}[E(Y|X_7)]}{V(Y)} = \frac{4}{70} = 0.06$$

$$S_8 = \frac{V_{X_8}[E(Y|X_8)]}{V(Y)} = \frac{25}{70} = 0.36.$$

The three least influential factors are X_1, X_5 and X_6, each one accounting for 1% of the output variance.

2. We discard the nonimportant factors from the model and we repeat the calculations with only the five remaining factors. We now have a new output variance:

$$V(Y) = \sum_{i=1}^{5} V(X_i) = 4+9+25+4+25 = 67,$$

and new elementary effects for the inputs:

$$S_1 = \frac{V_{X_1}[E(Y|X_1)]}{V(Y)} = \frac{4}{67} = 0.06$$

$$S_2 = \frac{V_{X_2}[E(Y|X_2)]}{V(Y)} = \frac{9}{67} = 0.14$$

EXERCISES

$$S_3 = \frac{V_{X_3}[E(Y|X_3)]}{V(Y)} = \frac{25}{67} = 0.37$$

$$S_4 = \frac{V_{X_4}[E(Y|X_4)]}{V(Y)} = \frac{4}{67} = 0.06$$

$$S_5 = \frac{V_{X_5}[E(Y|X_5)]}{V(Y)} = \frac{25}{67} = 0.37.$$

We see that the two most influential factors are X_3 and X_5, each one determining 37% of the output variance.

If we decide to fix those two factors at a given value in their range of variation, we will have only three factors varying, i.e. factors X_1, X_2 and X_4. In such a situation, the model will have a lower variance:

$$V(Y) = \sum_{i=1}^{3} V(X_i) = 4+9+4 = 17.$$

We conclude that, in this example, by fixing the two most important factors the output variance decreases from 76 to 17, with a reduction of 75%.

Exercise 5

1. Calculate the expansion of f into terms of increasing dimensionality (4.5) for the function (Ishigami and Homma, 1996):

$$f(X_1, X_2, X_3) = \sin X_1 + a \sin^2 X_2 + b X_3^4 \sin X_1. \qquad (4.34)$$

The input probability density functions are assumed as follows:[8]

$$p_i(X_i) = \frac{1}{2\pi},$$

when $-\pi \leq X_i \leq \pi$ and

$$p_i(X_i) = 0,$$

when $X_i < -\pi, X_i > \pi$ for $i = 1, 2, 3$.

---

[8] Note that this does not contradict the assumption that all factors are uniformly distributed within the unit hypercube Ω. It is always possible to map the hypercube to the desired distribution, and the sensitivity measure relative to the hypercube factors is identical to the measure for the transformed factors.

We calculate the decomposition of the function as (4.5) for $k = 3$:

$$f(X_1, X_2, X_3) = f_0 + f_1(X_1) + f_2(X_2) + f_3(X_3) + f_{12}(X_1, X_2)$$
$$+ f_{13}(X_1, X_3) + f_{23}(X_2, X_3) + f_{123}(X_1, X_2, X_3).$$

Thus

$$f_0 = E(Y) = \int \int \int f(X_1, X_2, X_3) p(X_1) p(X_2) p(X_3) dx_1 dx_2 dx_3$$

$$= \frac{1}{(2\pi)^3} \int \int \int (\sin X_1 + a \sin^2 X_2 + b X_3^4 \sin X_1) dx_1 dx_2 dx_3$$

$$= \frac{1}{(2\pi)^3} \left[\int \sin X_1 dx_1 + \int a \sin^2 X_2 dx_2 + \int \int b X_3^4 \sin X_1 dx_1 dx_3 \right]$$

$$= \ldots = \frac{a}{2}.$$

So $f_0 = a/2$.

The $f_i(X_i)$ terms are easily obtained:

$$f_1(X_1) = \int \int f(X_1, X_2, X_3) p(X_2) p(X_3) dx_2 dx_3 - f_0$$

$$= \frac{1}{(2\pi)^2} \left[(2\pi)^2 \sin X_1 + a \int \int \sin^2 X_2 dx_2 dx_3 \right.$$

$$\left. + b \sin X_1 \int \int X_3^4 dx_2 dx_3 \right] - \frac{a}{2}$$

$$= \ldots = \frac{1}{(2\pi)^2} \left[(2\pi)^2 \sin X_1 + 2a\pi^2 + \frac{4}{5} b \sin X_1 \pi^6 \right] - \frac{a}{2}$$

$$= \sin X_1 + \frac{1}{5} b \pi^4 \sin X_1 = \left(1 + \frac{1}{5} b \pi^4 \right) \sin X_1.$$

$$f_2(X_2) = \int \int f(X_1, X_2, X_3) p(X_1) p(X_3) dx_1 dx_3 - f_0$$

$$= \frac{1}{(2\pi)^2} \left[\int \int \sin X_1 dx_1 dx_3 + (2\pi)^2 a \sin^2 X_2 \right.$$

$$\left. + b \int \int X_3^4 \sin X_1 dx_1 dx_3 \right] - \frac{a}{2}$$

$$= \ldots = a \sin^2 X_2 - \frac{a}{2}.$$

$$f_3(X_3) = \int \int f(X_1, X_2, X_3) p(X_1) p(X_2) dx_1 dx_2 - f_0$$

$$= \frac{1}{(2\pi)^2} \left[\int \int \sin X_1 dx_1 dx_2 + a \int \int \sin X_2^2 dx_1 dx_2 \right.$$

EXERCISES

$$+bX_3^4 \int \int \sin X_1 dx_1 dx_2 \Big] - \frac{a}{2}$$

$$= \ldots = 0.$$

The $f_{ij}(X_i, X_j)$ terms are computed as

$$f_{12}(X_1, X_2) = \int f(X_1, X_2, X_3) p(X_3) dx_3 - f_1(X_1) - f_2(X_2) - f_0$$

$$= \sin X_1 + a \sin X_2^2 + b \sin X_1 \frac{1}{2\pi} \int X_3^4 dx_3 - f_1(X_1) - f_2(X_2)$$

$$- f_0 = 0.$$

$$f_{13}(X_1, X_3) = \int f(X_1, X_2, X_3) p(X_2) dx_2 - f_1(X_1) - f_3(X_3) - f_0$$

$$= \sin X_1 + a \frac{1}{2\pi} \int \sin X_2^2 dx_2 + bX_3^4 \sin X_1 - f_1(X_1) - f_3(X_3) - f_0$$

$$= \ldots = \left(bX_3^4 - \frac{1}{5} b\pi^4 \right) \sin X_1.$$

$$f_{23}(X_2, X_3) = \int f(X_1, X_2, X_3) p(X_1) dx_1 - f_2(X_2) - f_3(X_3) - f_0$$

$$= \frac{1}{2\pi} \int \sin X_1 dx_1 + a \sin X_2^2 + bX_3^4 \sin X_1 dx_1$$

$$- f_2(X_2) - f_3(X_3) - f_0$$

$$= 0$$

f_{123} is obtained by difference and is equal to zero.

2. Calculate the variances of the terms for the function, according to Equation (4.11).

First we calculate the unconditional variance of the function:

$$V(f(X)) = \int [f(X_1, X_2, X_3) - E(f(X))]^2 p(X_1) p(X_2) p(X_3) dx_1 dx_2 dx_3$$

$$= \frac{1}{(2\pi)^3} \int \int \int (\sin^2 X_1 + a^2 \sin^4 X_2 + b^2 X_3^8 \sin^2 X_1$$

$$+ 2a \sin X_1 \sin^2 X_2 + 2bX_3^4 \sin^2 X_1$$

$$+ 2abX_3^4 \sin X_1 \sin^2 X_2 + \frac{a}{4} dx_1 dx_2 dx_3$$

$$+ a \sin X_1 - a^2 \sin^2 X_2 - abX_3^4 \sin X_1)$$

$$= \frac{1}{(2\pi)^3} \left(\frac{1}{2} + \frac{3}{8} a^2 + \frac{b^2}{18} \pi^8 + \frac{a^2}{4} + \frac{b}{5} \pi^4 - \frac{a^2}{2} \right) (2\pi)^3$$

$$= \frac{1}{2} + \frac{a^2}{8} + \frac{b\pi^4}{5} + \frac{b^2 \pi^8}{18}.$$

We now calculate the variances of Equation (4.11), showing the passages for factor X_1:

$$V_1 = \int f_1^2(X_1)dx_1 = \int \left(\sin X_1 + \frac{1}{5}b\pi^4 \sin X_1\right)^2 dx_1$$

$$= \int \left[\sin X_1^2 + \frac{2}{5}b\pi^4 \sin X_1^2 + \frac{1}{25}b^2\pi^8 \sin X_1\right] dx_1 =$$

$$= \ldots = \frac{1}{2} + \frac{b\pi^4}{5} + \frac{b^2\pi^8}{50}.$$

The sensitivity index for factor X_1 can be calculated as

$$S_1 = \frac{V_1}{V} = \frac{1/2 + b\pi^4/5 + b^2\pi^8/50}{1/2 + a^2/8 + b\pi^4/5 + b^2\pi^8/18}$$

For the other factors we have

$$V_2 = \frac{a^2}{8}$$

$$V_3 = 0$$

$$V_{12} = 0$$

$$V_{13} = \frac{b^2\pi^4}{18} - \frac{b^2\pi^8}{50}$$

$$V_{23} = 0$$

$$V_{123} = 0$$

Again the fact that $V_{13} \neq 0$ even if $V_3 = 0$ is of particular interest, as it shows how an apparently noninfluential factor (i.e. with no main effect) may reveal itself to be influential through interacting with other parameters.

3. Show that the terms in the expansion of the function (4.34) are orthogonal.

Let us show, for example, that $f_1(X_1)$ is orthogonal to $f_2(X_2)$:

$$\int\int f_1(X_1)f_2(X_2)dx_1 dx_2 = \left(1 + \frac{1}{5}b\pi^4\right)\int\int \sin x_1 \left(a\sin^2 x_2 - \frac{a}{2}\right)$$

$$= \left(1 + \frac{1}{5}b\pi^4\right)\int \sin x_1 \int \left(a\sin^2 x_2 - \frac{a}{2}\right) dx_2,$$

which is equal to zero given that $\int \sin x_1 = 0$. The reader can verify, as a useful exercise, that the same holds for all other pairs of terms.

5

Factor Mapping and Metamodelling

With Peter Young

5.1 INTRODUCTION

WHERE WE DISCUSS ANOTHER CLASS OF QUESTIONS RELEVANT TO MODELLERS: HOW ARE THE MODEL OUTPUT VALUES PRODUCED? WHO IS MOST RESPONSIBLE TO DRIVE MODEL OUTPUT INTO SPECIFIC RANGES? CAN WE REPLACE THE ORIGINAL COMPLEX MODEL WITH A CHEAPER ONE, WHICH IS OPERATIONALLY EQUIVALENT?

In previous chapters we have dealt with sensitivity settings like Factor Prioritization (FP) and Factor Fixing (FF) and their associated methodologies. We saw in Chapter 4, for example, that variance-based techniques are able to provide unambiguous and rigorous answers to the questions posed in such settings; in particular, variance-based main effects suit factor prioritization, while total effects address the factor fixing setting. In the latter case, we also saw in Chapter 3 that elementary effect tests can provide excellent and cheap proxies for total effects, and can be used to screen models with a medium-to-large number of input factors. In Chapter 2, we discussed how to deal with models with very large numbers of input factors, exploiting sophisticated experimental designs.

We now turn to a different type of setting, which arises on foot of various questions often encountered in mathematical and computational modelling.

Modellers often need to address prototypical questions such as: 'Which factor or group of factors are most responsible for producing model outputs within or outside specified bounds? Which parameters determine uniqueness or instability or runaway conditions in a dynamic model?' For example, if Y were a dose of contaminant, we might be interested in how much (how often) a threshold level for this contaminant is being exceeded; or Y might have to fulfil a set of constraints, based on the information available on observed systems. The latter situation is typical in calibration.

Another typical question is whether it is possible to represent in a direct way (graphically, analytically, etc.) the relationship between input factors and output $Y = f(X_1, \ldots, X_k)$. Computing Y usually requires solving systems of nonlinear differential equations and the relationship $f(\cdot)$ can only be evaluated numerically, its form remaining unknown. Sensitivity analysis techniques discussed in previous chapters allow for ranking the importance of the various input factors in terms of influence on the variation of Y. In addition, some sort of direct representation of $Y = f(X_1, \ldots, X_k)$ would make the model's properties even more transparent.

In this chapter we will discuss techniques that can help to provide answers to such questions. We assign all these methods to the Factor Mapping setting, in which specific points/portions of the model output realizations, or even the entire domain, are mapped backwards onto the space of the input factors.

5.2 MONTE CARLO FILTERING (MCF)

WHICH FACTOR OR GROUP OF FACTORS ARE MOST RESPONSIBLE FOR PRODUCING MODEL OUTPUTS WITHIN OR OUTSIDE SPECIFIED BOUNDS? WHICH PARAMETERS DETERMINE UNIQUENESS OR INSTABILITY OR RUNAWAY CONDITIONS IN A DYNAMIC MODEL?

Let us first consider the case where the analyst is interested in targeted portions (extreme values, ceilings, thresholds, etc.) of the space of Y-realizations. In this situation, it is natural to partition the model realizations into 'good' and 'bad'. This leads very naturally to Monte Carlo filtering (MCF), in which one runs a Monte Carlo experiment producing realizations of the output(s) of interest corresponding to different sampled points in the input factors' space. Having done this, one 'filters' the realizations, i.e. the elements of the Monte Carlo sample that fall within the 'good' realizations are flagged as 'behavioural', while the remaining ones are flagged as 'nonbehavioural'. Regionalized Sensitivity Analysis (RSA, see Young *et al.*, 1978; Hornberger and Spear, 1981; Spear *et al.*, 1994; Young *et al.* 1996; Young 1999a and references cited therein) is an MCF procedure that aims to identify which factors are most important in leading to realizations of Y that are either in the 'behavioural' or 'nonbehavioural' regions. In typical cases, RSA can answer this question by examining, for each factor, the subsets corresponding to 'behavioural' and 'nonbehavioural' realizations. It is intuitive that, if the two subsets are dissimilar to one another (as well as, one would expect, to the initial marginal distribution of the factor), then that factor is influential.

5.2.1 Implementation of Monte Carlo Filtering

In Monte Carlo filtering a multiparameter Monte Carlo simulation is performed, sampling model parameters $(X_1, \ldots, X_k)$ from prior ranges and propagating parameter values through the model. Then, based on a set of constraints targeting the desired characteristics, a categorization is defined for each MC model realization, as either within or outside the target region. The terms behavioural (B) or nonbehavioural $(\bar{B})$ are current in the literature.

The $[B - \bar{B}]$ categorization is mapped back onto the input's structural parameters, each of which is thus also partitioned into a B and $\bar{B}$ subsample. Given a full set of N Monte Carlo runs, one obtains two subsets: $(X_i|B)$ of size n and $(X_i|\bar{B})$ of size $\bar{n}$, where $n + \bar{n} = N$. In general, the two subsamples will come from different unknown probability density functions, $f_n(X_i|B)$ and $f_{\bar{n}}(X_i|\bar{B})$.

In order to identify the parameters that are most responsible for driving the model into the target behaviour, the distributions f_n and $f_{\bar{n}}$ are compared for each parameter. If for a given parameter X_i the two distributions are significantly different, then X_i is a key factor in driving the model's behaviour and there will be clearly identifiable subsets of values in its predefined range that are more likely to fall under B than under $\bar{B}$. If the two distributions are not significantly different, then X_i is unimportant and any value in its predefined range is likely to fall into either B or $\bar{B}$.

This comparison can be made by applying standard statistical tests, such as the Smirnov two-sample test (two-sided version). In the Smirnov test the $d_{n,\bar{n}}$ statistic is defined for the cumulative distribution functions of X_i by

$$d_{n,\bar{n}}(X_i) = \sup \| F_n(X_i|B) - F_{\bar{n}}(X_i|\bar{B}) \|$$

and the question answered by the test is: 'At what significance level α does the computed value of $d_{n,\bar{n}}$ determine the rejection of the null hypothesis $f_n(X_i|B) = f_{\bar{n}}(X_i|\bar{B})$?'

The smaller α (or equivalently the larger $d_{n,\bar{n}}$), the more important the parameter is in driving the behaviour of the model. The procedure is exemplified in Figure 5.1 for a parameter X_i, uniformly distributed in the range $(0, 1)$ and displaying a significant difference between the B and $\bar{B}$ subsets. In order to identify the portion of X_i values more likely to fall

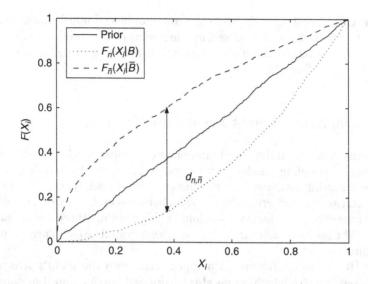

Figure 5.1 Graphical representation of the Smirnov test for an important parameter X_i. The vertical bar is the $d_{n,\bar{n}}$ statistic. Model realizations under the B category are more likely when X_i falls on the right of its predefined range

under B, the shape of the cumulative distribution $F_n(X_i|B)$ has to be examined. The latter is steeper on the right-hand side of the graph (or, equivalently, it has shifted downwards), indicating that X_i values on the right of its predefined range are more likely to produce a behavioural model realization.

The B and $\bar{B}$ subsets can be further probed through bidimensional projections, in order to detect significant patterns. The standard procedure consists of computing the correlation coefficients ρ_{ij} between all parameters under the B or $\bar{B}$ subsets, and plotting the bidimensional projections of the sample for the couples having $|\rho_{ij}|$ larger than a significance threshold. This usually makes it possible to 'visualize' relationships between parameters.

For example, let us consider a simple model given by the equation $Y = X_1 + X_2$, with $X_i \in (0, 1)$. Let us define the model's target behaviour as $Y > 1$. Then, an MCF procedure can identify a significant negative correlation between X_1 and X_2 in the B subset, and the corresponding triangular pattern can be visualized through the projection of the B sample shown in Figure 5.2. From this pattern, one can deduce a constraint $X_1 + X_2 > 1$ to fulfil the target behaviour.

The same procedure can evidently be applied in more typical cases where the constraint on the factors is not evident from the form of the mathematical model, i.e. when the model is a computer code.

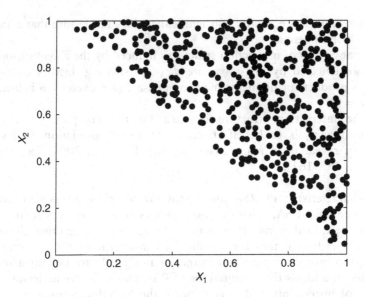

Figure 5.2 Bidimensional projection of the B subsample for the simple model $Y = X_1 + X_2$ (target behaviour $Y > 1$). The triangular pattern clearly indicates the relationship $X_1 + X_2 > 1$ for the target behaviour. The correlation coefficient between X_1 and X_2 in the MC sample is $\rho_{12} = -0.45$

5.2.2 Pros and Cons

Similarly to variance-based methods, RSA has many global properties: (a) the whole range of values of the input factors is considered, and (b) all factors are varied at the same time. Smirnov analysis considers univariate marginal distributions and it relates not only to main effects of variance-based methods, but can also highlight certain types of interaction effects (see the Exercises below). Higher-order analysis can only be performed for two-way interactions that are detectable through correlation analysis, but no procedure is provided for more complex interaction structure. Spear et al. (1994), reviewing their experience with RSA, highlighted two key drawbacks:

1. A low success rate: practice has shown that the fraction of B is barely larger than 5% over the total simulations for large models (with number of factors $k > 20$), implying a lack of statistical power;
2. Difficulty in uncovering correlation and interaction structures of the B subset (see also Beck's review, 1987):
   - the Smirnov test is sufficient to ascertain whether a factor under analysis is important. However, it does not provide a necessary condition

for importance, i.e. its nonsignificance does not ensure that a factor is noninfluential;
- many types of interaction structures induced by the classification are not detected by the univariate $d_{n,\bar{n}}$ statistic: e.g. factors combined as products or quotients may compensate (see Exercise 6 below, for $c = 0$);
- the interaction structure is often far too complex for correlation analysis to be effective, i.e. bivariate correlation analysis is not revealing in many cases (see Saltelli et al., 2004, Example 2, pp. 159–161).

Such characteristics of RSA imply that no complete assessment can be performed with RSA, since for those factors proving unimportant in the Smirnov test, further inspection is needed (e.g. by applying other global SA tools) to verify that they are not involved in interactions. Only after this subsequent inspection can the relevance of an input factor be fully assessed. In order to address these limitations of RSA and to better understand the impact of uncertainty and interaction in the high-dimensional parameter spaces of models, Spear et al. (1994) developed the computer-intensive Tree-Structured Density Estimation technique (TSDE), which allows for the characterization of complex interactions in that portion of the parameter space which gives rise to successful simulations. In TSDE, the B subsample is analysed by clustering regions of input factors characterized by high point density. This is based on a sequence of recursive binary splits of the B sample into two subdomains (similarly to peaks and tails of histograms) of complementary characteristics:

- small regions of relatively high density;
- larger sparsely populated regions.

The TSDE procedure relies on the assumption that any nonrandom density pattern indicates an influence of input factors on the model output. Interesting applications of TSDE in environmental sciences can be found in Spear (1997), Grieb et al. (1999) and Pappenberger et al. (2006). In the latter reference it is shown how factor mapping can be used to identify areas of desirable and undesirable model behaviour, which aids the modelling process. Helton et al. (2006) also provide mapping techniques on scatterplots, based on the same assumptions as TSDE.

Our experience suggests that such extended RSA techniques for mapping B subsets can be revealing *when they work*, but, in spite of their higher coding and computational complexity, they may still be characterized by lack of statistical power in discriminating the significance of such density patterns in a robust manner. Hence, we limit our discussion here to the Smirnov test and correlation analysis, due to the simplicity of their implementation and their ease of interpretation. Later on we will present more powerful mapping techniques when discussing metamodelling.

5.2.3 Exercises

1. Interpret the plots below, which represent the Smirnov test for a set of input factors. The behavioural set is indicated by dotted lines, the nonbehavioural by solid lines. Think of some functional forms that could produce such results. The D-stat above each plot indicates the value of the Smirnov statistic.

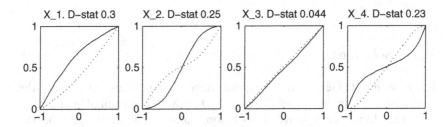

2. Interpret the plots below, showing the bidimensional projections of behavioural subsets. Think of an analytic form of the types of interaction that produce the behavioural sets.

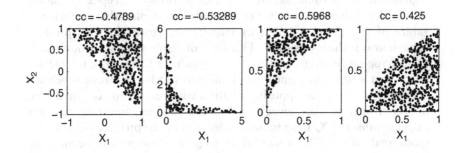

3. Consider the model $Y = Z_1 + Z_2$, with

$$Z_1 \sim N(0, 1)$$
$$Z_2 \sim N(0, 3).$$

Discuss the mapping problem $Y > 0$ analytically and using the MCF techniques.

4. For the same model as in Exercise 3, map the highest 5% quantile for Y onto the input space.
5. For the same model as in Exercise 3, map the highest 5% quantile for Y when

$$Z_1 \sim N(0, 1)$$
$$Z_2 \sim N(0, 1).$$

6. Consider the model $Y = X_1 \cdot X_2$, with $X_i \sim N(0, 2)$ and discuss the cases $Y > -1$, $Y > 0$, $Y > 1$.
7. Consider the model $Y = X_1 \cdot X_2 \cdot X_3$, with $X_i \sim U(0, 2)$ and discuss the case $Y < 1$.

5.2.4 Solutions

1. Three input factors out of four have a significant effect on the behavioural properties of the model. Only X_3 has a negligible effect.

    The dotted cumulative distribution curve for X_1 is steepest on the right-hand side (it has shifted downwards), so high values of X_1 are more likely to produce behavioural model realizations. This kind of result suggests a monotonic mapping between X_1 and Y, e.g. a simple linear relationship.

    Input factor X_2, on the other hand, has two different behavioural regions: the dotted line has two separate sections of steepness, one for the smallest values and one for the largest. This implies that the extreme values of X_2 (either smallest or largest) are more likely to produce behavioural realizations of Y. This kind of result suggests a nonmonotonic mapping between X_2 and Y, e.g. a quadratic form. Also, interaction effects can lead to the same type of Smirnov test (see Exercise 6 below).

    For X_4 we have the opposite situation with respect to X_2: the dotted line is steeper in the central part of the support, implying that the extreme values of X_4 have to be avoided in order to produce behavioural model realizations. This kind of result suggests a nonmonotonic mapping between X_4 and Y, e.g. a quadratic form with *opposite* concavity with respect to X_2.[1]
2. The first shape is characterized by a negative correlation between X_1 and X_2 in producing behavioural model realizations. Negative correlations suggest that the two input factors act through sum or product

---

[1] The true model used for this example was $Y = 4X_1 + (4X_2^2 - 1) - X_3^3 - (2X_4^4 - 1)$ with $X_i \sim U[-1, 1]$.

relationships. In this case, a plausible description for the behavioural set is given by $X_1 + X_2 > 0$, i.e. an additive relationship.

The second shape is also characterized by a negative correlation, suggesting action through sum/product. In this case, the shape resembles a hyperbola, suggesting a plausible functional form as $X_1 \cdot X_2 < const$.

The third shape is characterized by a positive correlation, implying that the action is now through difference/quotient. Possible relationships are $a \cdot X_1 - b \cdot X_2^2 < 0$ or $X_1 / X_2^2 < const$.

The fourth shape is again with positive correlation, but flipped with respect to the previous one, so plausible relationships are $a \cdot X_1 - b \cdot X_2^2 > 0$ or $X_1 / X_2^2 > const$.

3. The behavioural criterion $Y > 0$ is fulfilled by parameter combinations in the upper-right half plane delimited by the line $Z_1 = -Z_2$ (i.e. $Z_1 > -Z_2$). If we also consider the input factor distributions (Gaussian), we know that normal samples will fall into the range $\pm 1.96 \cdot \sigma$ with 95% probability, where σ is the standard deviation of the Gaussian distribution. So, the behavioural samples will be mainly concentrated (with 95% probability) in the upper part of an ellipse with vertical major axis of height 5.88 and horizontal minor axis of width 1.96, cut by the line $Z_1 = -Z_2$. Moreover, since Z_2 has a wider variance than Z_1, it will also be clear that Z_2 drives the sign of Y more powerfully, i.e. extreme values of Z_2 will be able to drive the sign of Y regardless of the actual values of Z_1.

We now perform the analysis applying the MCF approach (we give MATLAB commands as an example).

(a) Generate a sample of 1000 elements from two normal distributions of standard deviation 1 and 3:

```
x1 = randn(1000,1);

x2 = randn(1000,1).*3;
```

(b) generate the output:

```
y = x1+x2;
```

(c) look for behavioural elements of the sample:

```
ib = find(y>0);
```

and for the nonbehavioural:

```
in = find(y<=0);
```

(d) compute the Smirnov statistics (e.g. the outputs d1 and d2 of the MATLAB Statistical Toolbox function `kstest2`):

```
[h1, p1, d1] = kstest2(x1(ib),x1(in));

[h2, p2, d2] = kstest2(x2(ib),x2(in));
```

(e) plot the empirical cumulative density plots (e.g. using MATLAB Statistical Toolbox function `cdfplot`):

```
figure,

subplot(2,2,1)

h=cdfplot(x1(ib));

set(h,'linestyle',':'),

hold on, cdfplot(x1(in))

gca, title(['d-stat ',num2str(d1)]), xlabel('Z1'),
ylabel('')

subplot(2,2,2)

h=cdfplot(x2(ib));

set(h,'linestyle',':'),

hold on, cdfplot(x2(in))

gca, title(['d-stat ',num2str(d2)]), xlabel('Z2'),
ylabel('')
```

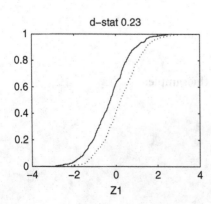
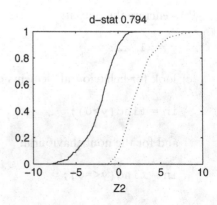

From the Smirnov analysis we can see that, while both input factors have a significant effect on the behavioural realizations of Y, Z_2 has the greater impact on the sign of Y (it has a far larger Smirnov statistic). As already mentioned, this is due to the larger variance of Z_2 with respect to Z_1, which allows sufficiently large values of Z_2 to force a positive sign in Y, regardless of the values of Z_1. The behavioural/nonbehavioural subsets for Z_2 are therefore almost disjoint (they overlap only in the range $[-1.5, 1.5]$ within a full support of $[-9, 9]$), while for Z_1 the two subsets have a much larger degree of overlap.

(f) compute the correlation coefficient under the behavioural subset:

```
cc = corrcoef(x1(ib),x2(ib))
```

and plot the bidimensional projection of the behavioural sample, which fills the half plane $Z_1 + Z_2 > 0$, as expected.

```
plot(x1(ib),x2(ib),'.')

xlabel('z1'), ylabel('z2'),
title(['cc=',num2str(cc(2,1))])
```

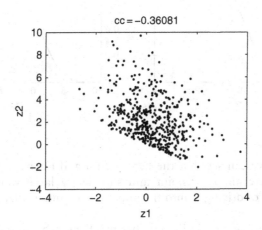

4. The output Y is the sum of two normally distributed variables, so $Y \sim N(0, \sqrt{10})$. The 5% upper tail of a Gaussian distribution is located at a distance from the mean of $1.65 \cdot \sigma$. Hence the upper 5% tail of the output probability is given by the set $Y > Y^{95}$, where $Y^{95} = 1.65 \cdot \sqrt{10} = 5.22$. So the behavioural set of the input factors is given by the upper half plane delimited by the line $Z_1 + Z_2 = Y^{95}$.

Performing the same analysis applying MCF techniques requires the use of the same sample used in Exercise 3 and recomputing the new behavioural set.

(a) Sort the output values:

```
[ys, is]=sort(y);
```

(b) define the behavioural (upper 5%) and nonbehavioural sets (the rest of the sample)

```
ib = is(951:1000);

in = is(1:950);
```

(c) compute the Smirnov statistics and plot the cumulative distributions as in the previous exercise:

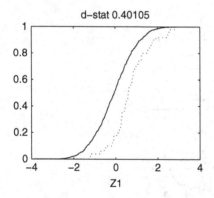
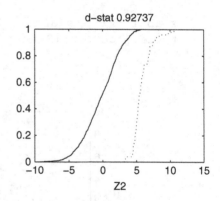

Again we can see that the behavioural and nonbehavioural subsets of Z_2 are almost disjoint and sufficiently large values of Z_2 are capable of driving Y into the upper 5% quantile, whatever the value of Z_1.

(d) compute the correlation coefficient of the behavioural set and plot the bidimensional projection of the input factor behavioural sample, which fills the half plane $Z_1 + Z_2 > Y^{95}$, as expected. This plot also confirms the Smirnov analysis by showing that Z_2 values have to remain significantly positive to drive Y to its upper values, while Z_1 values can range almost symmetrically around zero (see scatterplot on top of next page).

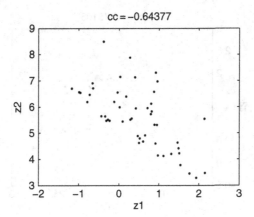

cc = −0.64377

5. In this case, the two input factors have the same variance, so we can expect an equal impact on the extreme values of Y. In analytic terms, Y now has a Gaussian distribution $N(0, \sqrt{2})$. Hence the upper 5% quantile is given by $Y > Y^{95} = 1.65 \cdot \sqrt{2} = 2.33$.

Performing the MCF analysis, we first obtain the Smirnov statistics and plot the cumulative distributions, which clearly display the similarity of the effects of the two input factors on Y.

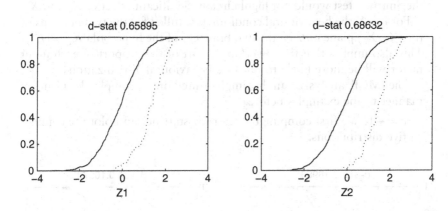

We can subsequently perform the correlation analysis and plot the bidimensional projection of the behavioural sample, which also confirms the equivalent effect of Z_1 and Z_2 on Y (see scatterplot on top of next page).

6. Analytically, the problems can be formalized as $Y > c$, with $c = -1, 0, 1$. This implies that $X_1 \cdot X_2 > c$.

For $c = -1$, the behavioural condition is fulfilled for the portion of the (X_1, X_2) plane *between* the two branches of the hyperbola $X_1 = -1/X_2$.

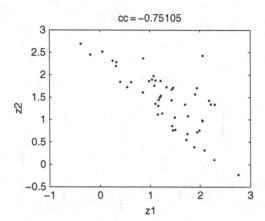

This also implies that the central part of the support of each input factor will be more likely to produce behavioural Y-realizations.

For $c = 0$, the behavioural condition is fulfilled if X_1 and X_2 have the same sign, i.e. all positive values of both X_1 and X_2 or all negative values of both X_1 and X_2. This also implies that any value in the support of one input factor has an equal probability of producing a behavioural or nonbehavioural run, conditional on the value of the other one. Therefore, the Smirnov test would not highlight any significant effect of X_1 and X_2.

For $c = 1$, the behavioural condition is fulfilled for the two portions of the (X_1, X_2) plane *outside* the two branches of the hyperbola $X_1 = 1/X_2$. This also implies that the lower/upper part of the support of each input factor will be more likely to produce behavioural Y-realizations.

The MCF analysis can be implemented in a completely identical manner to the examples before.

- $c = -1$. We first compute the Smirnov statistics and plot the cumulative distributions.

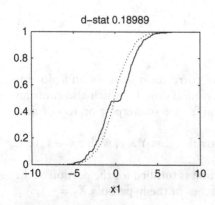
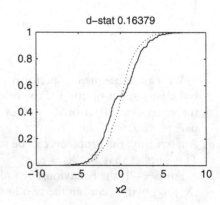

MONTE CARLO FILTERING (MCF)

Analysing the steepness of the cumulative distributions under B and $\bar{B}$ shows that the behavioural distribution is concentrated (steeper) in the central part of the initial support, while the nonbehavioural is concentrated (steeper) in two disjoint subsets in the lower and upper part of the initial support.

We then compute the correlation coefficient and plot the bidimensional projection of the behavioural subset.

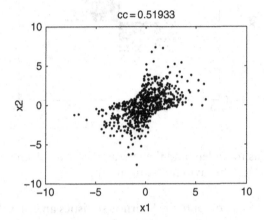

This shows neatly the portion of space between the two branches of hyperbola $X_1 = -1/X_2$ that produces behavioural Y-realizations.

- $c = 0$. We first compute the Smirnov statistics and plot the cumulative distributions.

This shows that the two subsets have the same distribution, i.e. any value in the original support for each input factor is equally likely to produce behavioural or nonbehavioural realizations.

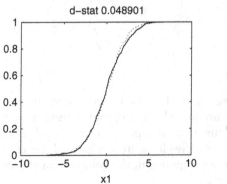
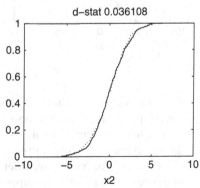

We then compute the correlation coefficient and plot the bidimensional projection of the behavioural subset.

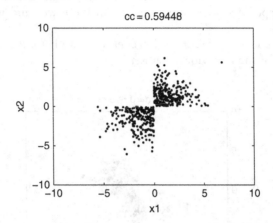

This allows us to understand the interaction mechanism between X_1 and X_2 that produces behavioural realizations.

- $c = 1$. We first compute the Smirnov statistics and plot the cumulative distributions.

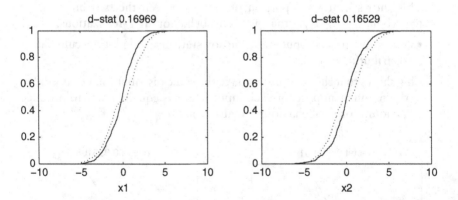

This shows that values of the input factors concentrated in the lower/upper part of each support are more likely to produce behavioural runs, while the central values of the original support are excluded from the behavioural set (the dotted cumulative lines are flat around the zero values of Z_i, implying a zero density of points of the behavioural subset in the central part of each support).

We then compute the correlation coefficient and plot the bidimensional projection of the behavioural subset.

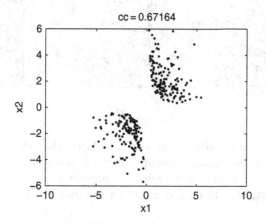

This shows the two disjoint portions of the (X_1, X_2) plane outside the two branches of the hyperbola $X_1 = 1/X_2$ that produce the behavioural Y-realizations.

7. We first compute the Smirnov statistics and plot the cumulative distributions.

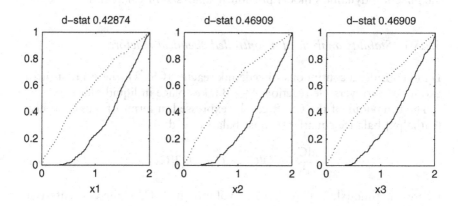

This shows that, to maintain the output realization smaller than 1, all factors must be sufficiently small. The behavioural distribution is therefore steepest towards the smallest values of the original supports of X_i.

We then compute the correlation coefficient and plot the bidimensional projection of the behavioural subset.

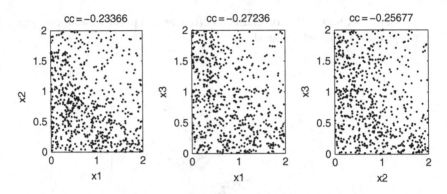

The negative correlation coefficients indicate that the effect of input factors on Y is through sums/products. Moreover, these plots show nicely that the upper-right regions, combining large values of all the input factors, have to be avoided to ensure that $Y < 1$.

5.2.5 Examples

We now show a few examples of the use of MCF to characterize the stability behaviour of dynamic models. We will give an example for a chemical reactor (continuous time model) and for a macroeconomic model (discrete time model). Finally, we will also consider, in the light of MCF techniques, the infection dynamics model previously analysed in Chapter 4.

5.2.5.1 *Stability analysis of a controlled chemical reactor*

Let us consider a continuous stirred tank reactor (CSTR) where a first-order exothermic irreversible reaction $A \to B$ takes place in liquid phase.

The behaviour of the CSTR can be expressed in terms of mass and heat (enthalpy) balance equations. Mass balance reads as

$$V \frac{dC_A}{d\bar{t}} = Q(C_{A0} - C_A) - k(T)C_A \cdot V \qquad (5.1)$$

where $\bar{t}$ is time [s], V is the reactor volume [m³]; C_A is the concentration of A in the reactor and at its outlet [kmol/m³]; C_{A0} is the concentration of A at the inlet; Q is the volumetric flow rate [m³/s] at the input and output of the reactor, $k(T)$ is the kinetic 'constant' [1/s] of the first-order chemical reaction $A \to B$, which is expressed as a function of temperature T [K]:

$$k(T) = k_0 \exp\left(-\frac{E}{RT}\right),$$

where k_0 is the Arrhenius factor [1/s], E is the activation energy [kJ/kmol] and R is the gas constant [kJ/(kmol K)].

The mass balance equation (5.1) tells us that the rate of change of the amount of reactant A, given by the left-hand side term $V(\mathrm{d}C_A/\mathrm{d}\bar{t})$, equals the flow of A at the inlet, $Q \cdot C_{A0}$, minus the flow of A at the outlet, $Q \cdot C_A$, and minus the amount of A that is transformed into B per time unit, $k(T) \cdot C_A V$.

The heat balance reads as

$$V\rho c_p \frac{\mathrm{d}T}{\mathrm{d}\bar{t}} = Q\rho c_p(T_0 - T) - \Delta H_r k(T) C_A V - UA(T - T_c) \quad (5.2)$$

where ρ is the density of the reacting mixture [kg/m^3], c_p is the specific heat of the mixture [kJ/(kg K)], T_0 is the temperature of liquid entering the reactor [K], T is the temperature of the liquid in the reactor and at its outlet, T_c is the temperature of the reactor's coolant, $(-\Delta H_r)$ is the reaction enthalpy [kJ/kmol], U is the overall heat transfer coefficient between the inside of the reactor and the coolant [kJ/(s m^2 K)] and A is the heat transfer area [m^2].

The heat balance equation (5.2) tells us that the rate of change of enthalpy in the reactor, $V\rho c_p(\mathrm{d}T/\mathrm{d}\bar{t})$, equals the flow of enthalpy at the inlet, $Q\rho c_p T_0$, minus the flow of enthalpy at the outlet, $Q\rho c_p T$, plus the heat generated by the exothermic reaction, $(-\Delta H_r) k(T) C_A V$, minus the heat removed by the coolant, $UA(T - T_c)$.

The CSTR is controlled, in order to keep the temperature, and the associated quantity of product B, at the desired set-point T_s. The controlled variable is the temperature T and the manipulated variable is the coolant temperature T_c. The controller follows a standard proportional-integral (PI) design, implying the following control rule for the coolant temperature:

$$(T_c - T_{cs}) = -k_P(T - T_s) - k_I \int_0^t (T - T_s) \mathrm{d}\bar{t} \quad (5.3)$$

where T_{cs} is the coolant temperature at the set-point, k_P is the proportional control gain and k_I is the integral control gain [1/s]. The control rule (5.3) tells us that the coolant 'error' $(T_c - T_{cs})$ is proportionally adjusted, with opposite sign, with respect to the temperature error $(T - T_s)$ and with respect to the integral of the temperature error. In other words, the coolant temperature will go down as the temperature error $(T - T_s)$ and its integral go up. The control gains k_P and k_I tell us by *how much* the coolant temperature is adjusted by the controller for a given temperature error and integral of the error: the higher the gains, the greater the change in the coolant. Finally, the integral action ensures that the desired set-point is the unique steady state of the reactor.

Economists or econometricians will be familiar with Taylor rules: in monetary policy, for example, the controlled variable is inflation and the manipulated variable is the nominal interest rate. When the economy is 'overheated' (i.e. with high inflation), the Central Bank *increases* the nominal interest rate to 'cool' the economy. In the case of the CSTR, the PI regulator *decreases* coolant temperature to cool an overheated reactor.

The controlled CSTR is therefore described by the following system of differential equations, in dimensionless form:

$$\frac{d\xi}{dt} = -\xi + Da(\vartheta)(1-\xi) \tag{5.4}$$

$$\frac{d\vartheta}{dt} = \vartheta_0 + N[\vartheta_{cs} - k_P(\vartheta - \vartheta_s) - k_I\tau\varphi] - (1+N)\vartheta + Da(\vartheta)(1-\xi) \tag{5.5}$$

$$\frac{d\varphi}{dt} = \vartheta - \vartheta_s \tag{5.6}$$

where (notation of Pellegrini and Biardi, 1990)

$\xi = \frac{C_{A0} - C_A}{C_{A0}}$ (conversion of A)

$\tau = \frac{V}{Q}$ (residence time [s])

$t = \frac{\tilde{t}}{\tau}$ (dimensionless time)

$N = \frac{UA}{Q\rho c_p}$ (dimensionless heat transfer coefficient)

$\vartheta = \frac{T - T_s}{\Delta T_a}$ (dimensionless temperature)

$\Delta T_a = \frac{-\Delta H_r C_{A0}}{\rho c_p}$ (adiabatic temperature rise [K])

$Da(\vartheta) = k_0 \tau \exp(-\frac{E}{R(T_s + \Delta T_a \vartheta)})$ (Damkoehler number)

$\varphi = \int_0^t (\vartheta - \vartheta_s) dt$ (dimensionless integral of the error)

The adiabatic temperature rise indicates the temperature increase that would be caused in the reactor if the entire amount of input A were converted into B under adiabatic conditions (i.e. without any heat exchange). The Damkoehler number indicates the average number of 'reaction events' that occur during the residence time.

The behaviour of this dynamical system, which can present complex dynamic features, from instability of the steady state to chaotic behaviour, has been intensively studied (see Pellegrini and Biardi, 1990; Giona and Paladino, 1994; Paladino *et al.*, 1995; Paladino and Ratto, 2000).

Our aim in the present example is to study the stability conditions of the controlled CSTR. The local stability analysis of the reactor in the neighbourhood of the unique steady state $(\xi_s, \vartheta_s, \varphi_s)$ is performed by analysing the Jacobian:

$$J = \begin{pmatrix} -(1+a_{21}) & a_{12} & 0 \\ -a_{21} & -(k_P N + N + 1) + a_{12} & -k_I \tau \\ 0 & 1 & 0 \end{pmatrix} \tag{5.7}$$

where

$$a_{12} = \xi_s \frac{E\Delta T_a}{R(T_s + \Delta T_a \vartheta_s)^2} > 0 \tag{5.8}$$

$$a_{21} = Da(\vartheta_s) > 0 \tag{5.9}$$

and $\xi_s = Da(\vartheta_s)/(1 + Da(\vartheta_s))$.

The steady state is stable if all three eigenvalues of J have negative real parts. This ensures that, as the operating conditions are moved away from the set-point (i.e. the steady state), the reactor will return to steady state. Many authors have demonstrated that this system presents a Hopf bifurcation locus. At the Hopf locus the steady state becomes unstable and the dynamic behaviour of the reactor is characterized by persistent oscillations (limit cycle). This is, of course, unacceptable and must be avoided.

Although the Hopf bifurcation locus can be computed analytically (Giona and Paladino, 1994), here we analyse the stability conditions by applying the MCF techniques described in this chapter. This will allow us to confront the results of the MCF analysis with the analytic results. The problem can be formalized in the MCF framework as follows:

- the input factors are the control gains and the uncertain physicochemical parameters of the CSTR model;
- the outputs are the eigenvalues of the Jacobian;
- the filtering criterion is:

    - behaviour B if all eigenvalues have negative real parts;
    - nonbehaviour $\bar{B}$ otherwise.

The nominal conditions of the CSTR are defined as follows:

$$
\begin{array}{llll}
k_0 = 133600s^{-1} & E/R = 8000K & \tau = 3600s & \Delta T_a = 200K \\
T_0 = 298.42K & T_s = 430K & T_{cs} = 373.16 & \\
\vartheta_s = 0 & \xi_s = 0.8 & N = 0.5 &
\end{array}
\tag{5.10}
$$

Such nominal conditions, depending on the values of the various physicochemical parameters, are subject to a degree of uncertainty.

Let us first analyse the stability of this system under the nominal conditions, by varying only the control gains k_P and $k_I \tau$. As anticipated, this analysis can be performed analytically. We analyse here the Hopf bifurcation locus in the $(k_P, k_I \tau)$ plane using the MCF approach. The analysis requires the following steps:

- Sample the control gains uniformly in the range [0, 10];
- Compute the eigenvalues of the Jacobian;
- Check the stability condition;

  - the set of control gains providing stable eigenvalues (negative real parts) is the behavioural set;
  - the set of control gains providing unstable eigenvalues (nonnegative real parts) is the nonbehavioural set;

- Perform the Smirnov analysis;

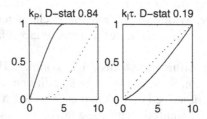

Looking at the plot for k_P, we can see that the nonbehavioural cumulative distribution (solid line) has a limit threshold at about $k_p^* = 5$, above which only stable solutions are present; this implies that sufficiently large values of the proportional control gain ($k_P > k_P^*$) are able to stabilize the reactor, whatever the value of the integral control gain. The latter gain, on the other hand, is more likely to produce a stable reactor for small values. However, the two cumulative distributions for $k_I \tau$ have the same support, i.e. both ranges of the stable and unstable sets span the entire support [0, 10], implying that no clear threshold of stability can be identified.
- Plot the behavioural sample on the $(k_P, k_I \tau)$ plane.
  This shows the boundary of stability that exactly corresponds to the analytic solution (solid line). Note also that the patterns in the scatterplot are due to the Sobol' quasi-random sequences used for the example.

We now check the stability analysis for robustness, by allowing physico-chemical parameters to be uncertain. These uncertainties are given by normal distributions, with the following characteristics:

- $(k_0, E/R)$: assuming an estimate of kinetic coefficients, they are likely to be strongly correlated (Bard, 1974) and with much greater uncertainty for k_0 than for E/R. So we take a standard deviation of 35% for k_0,

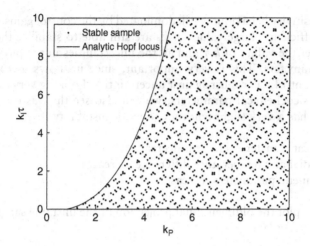

2% for E/R and a correlation coefficient of 0.96 (Paladino and Ratto, 2000).
- (N): the heat transfer coefficient has a standard deviation of 5%, i.e. $\sim N(0.5, 0.025)$.
- (ΔT_a): the adiabatic temperature difference has a standard deviation of 5 K, i.e. $\sim N(200, 5)$.

We then proceed with the MCF analysis, as described below.

- First we perform the Smirnov test.

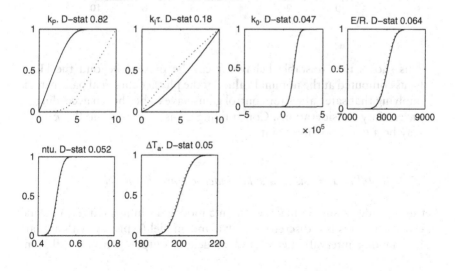

The sensitivity behaviour is still dominated by the control gains. In particular, sufficiently large values of k_P are still able to stabilize the reactor, whatever the values of $k_I \tau$ and all the uncertainties in the physicochemical parameters. This is very important, since it allows us to design a stable control, i.e. robust against uncertainties. Looking very carefully at the physicochemical parameters, we can also see that, as expected, the reactor has a very slight tendency towards instability if:

- the heat transfer coefficient decreases;
- the adiabatic temperature difference increases;
- the kinetic parameters increase.

- We then plot the bidimensional projection of the unstable sample $\bar{B}$ onto the $(k_P, k_I \tau)$ plane.

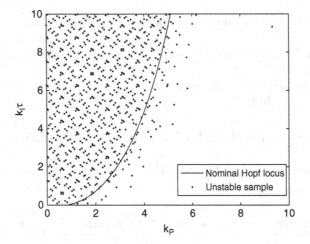

This shows that unstable behaviour can be present beyond the Hopf locus computed at the nominal values of the physicochemical parameters, implying that safe values for control gains have to be chosen according to uncertainty considerations. Constraining gains as $k_P > 6$ and $0 < k_I \tau < 4$ may be a good starting point.

5.2.5.2 Stability analysis of a small macroeconomic model

Let us consider a simple macroeconomic model: a Phillips curve. As usual in economics, this is a discrete-time dynamic model (typically with a quarterly sampling interval). Let c_t and π_t denote output gap and inflation, respectively. In macroeconomic theory, the output gap denotes the cyclical component of GDP (gross domestic product) with respect to the long-term trend. In general terms, it is a quantity linked to the business cycle. We can write the hybrid Phillips curve as

$$\pi_t = \omega_b \pi_{t-1} + \omega_f E_t \pi_{t+1} + \beta c_t + a_{\pi t} \tag{5.11}$$

$$c_t = 2A\cos(2\pi/\tau)c_{t-1} - A^2 c_{t-2} + a_{c,t} \tag{5.12}$$

where E_t denotes the expectation taken at period t, $0 < (\omega_b, \omega_f) < 1$, A and τ are the amplitude and period of c_t and $a_{\pi,t}$, $a_{c,t}$ are white noises.

The Phillips curve links the inflation dynamics to the output gap, in such a way that periods of economic expansion (i.e. with a positive output gap) are typically associated with an increase in inflation and vice versa. Moreover, the hybrid Phillips curve also says that inflation in the current period is linked with some persistence ω_b to the rate of inflation in the previous period and to the expected level of inflation in the following period, with a weight ω_f. This leads on to the theory of rational expectation behaviour of economic agents. In contrast to standard (physical) dynamic systems, the occurrence of a unique, stable solution in macroeconomic rational expectations models requires that there be an equal number of explosive eigenvalues and forward-looking variables. In discrete-time dynamic models, stable roots have absolute values less than 1, while explosive ones are larger than 1. In this case, the Phillips curve has one lag π_{t-1} and one lead $E_t \pi_{t+1}$, so we need exactly one stable and one explosive eigenvalue. To help explain the stability conditions of economic rational expectations models to non-economists, we can say that the fact that the current level of inflation depends on both past and future levels, makes the system like a two-point boundary system (similar to certain types of differential equations in space describing advection–dispersion mechanisms). Hence, this implies the presence of initial and terminal conditions, corresponding to backward-looking and forward-looking components, respectively. Likewise in physical systems, backward-looking behaviour propagates the initial conditions into the future. This propagation is stable if it is associated to stable eigenvalues, thus assuring that the dynamic system will asymptotically converge to the steady state. The forward-looking components, on the other hand, propagate the terminal conditions into the past, i.e. in a symmetrical manner, reversing the orientation of the time axis. It intuitively makes sense that, if the orientation of time is reversed, explosive roots looking 'towards the future' become stable roots looking 'towards the past'; therefore, in order to ensure stability of the propagation into the past of forward-looking components, such components need to be associated to explosive roots.

The eigenvalues of this simple model can be computed analytically:

$$r_b = \frac{1/\omega_f - \sqrt{1/\omega_f^2 - 4\omega_b/\omega_f}}{2}$$
$$= \frac{1 - \sqrt{1 - 4\omega_b\omega_f}}{2\omega_f} \quad (5.13)$$
$$r_f = \frac{1 + \sqrt{1 - 4\omega_b\omega_f}}{2\omega_f}.$$

The restriction $1 - 4\omega_b\omega_f \geq 0$ ensures that the roots are real. The stability condition is verified if

$$\omega_f < (1 - \omega_b) \quad (5.14)$$

or

$$\omega_f = (1 - \omega_b) \quad \text{and} \quad \omega_b > 0.5.$$

We apply the Monte Carlo filtering technique to identify the stable behaviour. The support for the model coefficients is defined as

$$A \sim U[0, 1], \quad \omega_b \sim U[0, 1], \quad \omega_f \sim U[0, 1], \quad \tau \sim U[0, 100].$$

We first perform the Smirnov test for the separation of the B and $\bar{B}$ subsets.

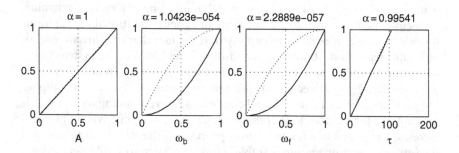

The occurrence of stable or unstable behaviour is clearly attributable to the coefficients ω_b and ω_f. Moreover, the shape of the dotted curves (corresponding to B) indicates that stable behaviour will more probably occur for smaller ω_b and ω_f values.

We subsequently perform a correlation analysis of the stable sample and plot the significant correlation selected: the bidimensional projection of the B subset onto the (ω_b, ω_f) plane.

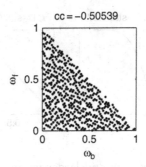

The dots in this plot clearly indicate the first stability condition $\omega_f < (1 - \omega_b)$ in (5.14). The second condition in (5.14) is just one limit case, and tells us simply that only the half part $\omega_b > 0.5$ of the stability boundary line $\omega_f = (1 - \omega_b)$ provides stable behaviour. This is, of course, hardly visible in the plot.

5.2.5.3 Mapping propagation of the infection in the simple infection dynamics model

Let us consider the model presented in Chapter 4. We want to map the conditions under which infection propagates. Recalling the solution to the model shown in Chapter 4, we have

$$I = I_0 \exp(Yt)$$

with

$$Y = \gamma k S_0 - r - d$$

which states that the infection dies out if $Y < 0$ and propagates if $Y > 0$. So, our MCF problem is to map $Y < 0$, i.e. the stable eigenvalue of the infection propagation dynamics.

We sample input factors from the same distributions used in Chapter 4 for the three different scenarios of parameter k.

- $k \sim beta(2, 7)$. In this case, the probability that the infection will die out is very small (only 1% of the MC sample). Performing the Smirnov analysis we obtain the picture on the top of next page.

  This shows that all input factors have a nonnegligible effect in driving the propagation of the infection. However, γ is predominant, based on which it is clear that the infection can die out only for a very narrow range of γ values, in the lowest part of its range. Moreover, we can also

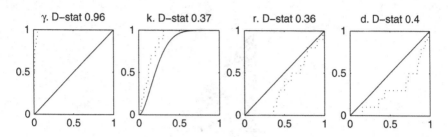

see that smaller values of k and larger values of r and d tend to limit propagation of the infection.

The correlation analysis reveals a negative correlation between γ and k under the behavioural subset, which reflects the product interaction between these two factors. Note also that γ-values in this behavioural scatterplot are constrained in the range $[0, 0.05]$ out of a full sample in the range $[0, 1]$.

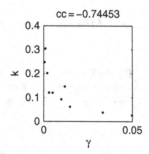

- $k \sim \text{beta}(0.5, 10)$. Changing the prior distribution of k raises the probability that the infection will die out to about 20%. Performing the Smirnov analysis produces the following modified picture for the model parameters:

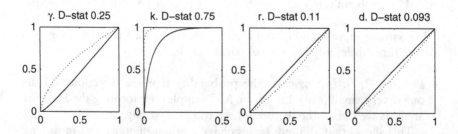

Now the importance of k has increased, and we can see that the behavioural runs are linked merely to k, falling in the lowest part of its range. Moreover, small γ values and large r and d values also tend to produce a declining infection dynamics.

The correlation analysis confirms the negative correlation between γ and k, while the hyperbola shape linked to the product interaction of these two parameters in the model is now more sharply visible.

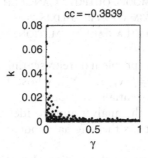

- $k \sim beta(0.2, 15)$. In this case, the probability of a declining infection dynamics rises to 57%. The Smirnov analysis indicates that k is now the dominant factor in driving the behaviour of the infection dynamics model, leaving a minor role to the remaining parameters:

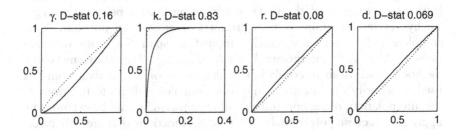

The correlation analysis still produces a negative correlation between γ and k with the hyperbola boundary between the B and $\bar{B}$ subsets.

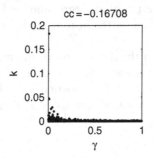

5.3 METAMODELLING AND THE HIGH-DIMENSIONAL MODEL REPRESENTATION

IS IT POSSIBLE TO REPRESENT IN A DIRECT WAY THE RELATIONSHIP BETWEEN INPUT FACTORS AND THE MODEL OUTPUT? CAN SUCH A DIRECT AND COMPUTATIONALLY CHEAPER APPROXIMATION BE USED IN PLACE OF THE ORIGINAL MODEL? HOW CAN WE ESTIMATE SUCH AN APPROXIMATION?

Let us now consider the problem of representing in a clear and immediate way the relationship $Y = f(X_1, \ldots, X_k)$, whose form is usually unknown to the analyst. This is essentially a problem of model approximation or metamodelling, whereby the analyst aims to identify a simple relationship between X_i's and Y that fits the original model well and is less computationally demanding.

There is a vast literature on this subject. Local approximation methods take the value of f and its derivatives at a base point $\mathbf{X}_0$ and construct a function that matches the properties of f at $\mathbf{X}_0$ and in the nearby region (Taylor series). Interpolation methods look at 'nice' functions that go through a set of data points spanning the entire domain of the $Y = f(\cdot)$ mapping. The approximation is then identified by fixing p parameters (e.g. the coefficients of the polynomials) using p data points (Lagrange, Chebyshev interpolation). Tensor products of orthogonal polynomials or complete polynomials are usually applied to span the space of functions in R^k and to interpolate $Y = f(X_1, \ldots, X_k)$. Regression/smoothing methods differ from interpolation in that a set of $N > p$ data points is used to identify the approximating function. For univariate functions f, the interpolation, regression and smoothing approaches can be extended by applying piecewise polynomials, constructing functions that are only piecewise smooth. Splines (cubic splines are the most popular) are a powerful and widely used approach to piecewise polynomial interpolation and regression/smoothing (in the latter case they are called smoothing splines). Splines are smooth where the polynomial pieces connect. In the multivariate case, radial basis function (RBF) networks can be seen as the equivalent of univariate piecewise interpolation, regression and smoothing approaches (RBFs are also classified under the heading of kernel regression smoothing methods).

Before proceeding with the description of the methodologies, we would like to draw attention to some additional properties of the approximating functions, which also uncover fundamental links between metamodelling and the theory of variance-based sensitivity analysis discussed in previous chapters.

Let $g(\cdot)$ be the generic function approximating the true model $Y = f(X_1, \ldots, X_k)$ and let us assume a quadratic loss function $E[(Y - g(\cdot))^2]$ as a measure of 'fit' for g.

If we were to approximate f with a function of *one single* parameter X_i, what function $g_i^*(X_i)$ would produce minimum loss?

It is well known, from any standard text on statistics, that the univariate function

$$g_i^* = E[Y|X_i], \quad (5.15)$$

i.e. the conditional expectation of Y given X_i, is the minimum loss approximation to f. The expression (5.15) tells us that at any p-location $X_i = x_{i,p}$, the value of g^* is obtained by integrating (averaging) Y over all the remaining $[X_1, \ldots, X_{i-1}, X_{i+1}, \ldots, X_k]$ input factors.

Equation (5.15) can be generalized to any subset of the input factors $\mathbf{X}_I$, indexed by $I = (i_1, \ldots, i_l)$, as

$$g_I^* = E[Y|\mathbf{X}_I]. \quad (5.16)$$

Equations (5.15) and (5.16) obviously link to the ANOVA-HDMR decomposition of f discussed in Chapter 4:

$$f(X_1, X_2, \ldots, X_k) = f_0 + \sum_i f_i + \sum_i \sum_{j>i} f_{ij} + \ldots + f_{12\ldots k}, \quad (5.17)$$

where the connection between the f_i terms of the HDMR and the minimum loss approximating functions $g^*(\cdot)$ is made explicit by

$$f_0 = E(Y)$$
$$f_i(X_i) = E(Y|X_i) - f_0 = g_i^* - f_0 \quad (5.18)$$
$$f_{ij}(X_i, X_j) = E(Y|X_i, X_j) - f_i(X_i) - f_j(X_j) - f_0 = g_{i,j}^* - f_i(X_i) - f_j(X_j) - f_0.$$

Each term of the ANOVA-HDMR decomposition tells the analyst how much, on average, Y can be moved with respect to its mean level f_0 by acting on single input factors or groups of them. Moreover, the quantity $V(g_I^*(\mathbf{X}_I))/V(Y) = \mathrm{corr}(g_I^*(\mathbf{X}_I), Y)$ is well known in statistics as the 'correlation ratio' or 'nonparametric R-squared', and provides the fraction of the variability of Y that is explained with the *best* predictor based on $\mathbf{X}_I$. The equivalence between nonparametric R-squared and variance-based sensitivity indices is obvious and this closes the parallel between the problem of estimating and measuring the explanatory power of covariates in regression and variance-based sensitivity analysis.

Coming back to the metamodelling problem, kernel regression methods can be shown to provide, under certain regularity conditions, consistent estimators of $g^*(\cdot)$, which are asymptotically normal at the $\sqrt{N}$ rate (see Doksum and Samarov, 1995, and references cited therein), i.e. as $N \to \infty$,

$$E \int (\hat{g}(\mathbf{X}_I) - g^*(\mathbf{X}_I))^2 d\mathbf{X}_I = o(N^{-1/2}).$$

The metamodelling approach that we follow in this book can be classified under kernel regression smoothing methods. However, due to its recursive implementation, our approach differs with respect to other *en-bloc* methods. This has some advantages, such as the estimation of 'smoothing parameters' ('hyper-parameters') with maximum likelihood and greater flexibility in managing nonlinearities in $f(\cdot)$.

5.3.1 Estimating HDMRs and Metamodels

In the literature on sensitivity analysis there has been a growing interest in metamodelling and smoothing techniques. Storlie and Helton (2008) have reviewed smoothing methods for sensitivity analysis, from smoothing splines to various types of univariate and multivariate kernel regression approaches. Li *et al*. (2002, 2006) developed the so-called Random Sampling HDMR, which involves approximating the truncated HDMR expansion up to order three, based on orthogonal polynomials. Pappenberger and Stauch (2007) use spline smoothing to estimate sensitivity indices. Using the State-Dependent Regression (SDR) approach of Young (2001), Ratto *et al*. (2004, 2007) have developed a nonparametric approach which is very similar to smoothing splines and kernel regression approaches, but which is based on recursive filtering and smoothing estimation (the Kalman Filter, KF, combined with Fixed Interval Smoothing, FIS). Such a recursive least-squares implementation has some fundamental advantages: (a) it is couched with optimal maximum likelihood estimation, thus allowing for an objective estimation of the smoothing hyper-parameters, and (b) it allows for greater flexibility in adapting to local discontinuities, heavy nonlinearity and heteroscedastic error terms (see below). All such methods can be assigned to the regression/smoothing class of approximation approaches.

An example of interpolating metamodels, on the other hand, is given by Gaussian emulators (see Oakley and O'Hagan, 2004, and the references cited therein, for a detailed description) and kriging metamodels (Kleijnen, 2007a,b). Kriging metamodels are similar to Gaussian, except that they do not rely on Bayesian interpretation. While *theoretically* appealing, Gaussian emulators can be prone to the curse of dimensionality and to the smoothness assumptions of the function under analysis. This is because, instead of trying to identify the best predictors of Y based on a subset of input factors or on low-order ANOVA-HDMR terms, Gaussian emulators try to interpolate and predict the $f(\cdot)$ mapping by applying a Gaussian kernel of the same k-dimensionality as the input parameter space. Therefore, as k increases, the number of 'hyperparameters' to be estimated (linked to the covariance structure of the k-dimensional Gaussian kernel), increases strongly, often implying problems with identification and overparameterization.

Such problems are well known in standard interpolation and smoothing techniques based on k-dimensional kernel regressors. They imply that Gaussian emulators are only effective, *in practice*, for model structures having a small number of significant main effects and very mild interactions, for which such problems are made irrelevant by the very few highly identifiable elements of the $f(\cdot)$ mapping.

In other words, in regression/smoothing techniques, metamodels are based on subsets of input factors and/or truncated HDMR expansions of order smaller than k, and their identification and estimation *incorporate* sensitivity analysis criteria, in that nonsignificant contributions to Y are identified and eliminated *within* the process of construction of the approximation to Y. In contrast to this, Gaussian emulators aim first to estimate a full k-order mapping on the basis that a sensitivity analysis applied afterwards to the emulator will automatically reveal the significant contributions to Y.

Once identified, estimated and parameterized, metamodels provide a direct, albeit approximated, analytic expression of the $Y = f(X_1, \ldots, X_k)$ mapping, which accounts for nonlinearities and interaction terms of increasing order. As such, they can be used for various purposes:

- sensitivity analysis, by helping to highlight the most important input factors of the mapping;
- model simplification, by finding a surrogate model containing a subset of the input factors that account for most of the variability of Y;
- model calibration, in which the metamodel is used to find directly the optimal parameterization for the fulfilment of the given calibration criteria.

A detailed description of all the available metamodelling techniques is beyond the scope of this book: readers can refer to the cited works for further information. We will concentrate here on nonparametric methods and, in particular, we will demonstrate simple implementations of univariate nonparametric smoothing methods that give the 'flavour' of the more sophisticated multivariate, recursive procedure of Young (2001), as used for metamodelling in Ratto *et al.* (2004, 2007). In nonparametric methods, the function $E(Y|X_i)$ is not approximated by a basis of functions that span the entire domain of X_i; rather many 'local' approximations are identified which move along the X_i-axis. Such 'local' functions are subsequently joined using ad hoc criteria, such as imposing some smoothness properties like continuity (piecewise linear interpolation is an example of this). This in practice gives a 'look-up' table of the function $f_i = g^*(X_i)$ which can subsequently be parameterized using functional bases, such as polynomials, Fourier expansions, linear wavelets or Radial Basis functions (RBFs).

5.3.1.1 Smoothing scatterplots using the Haar wavelet

Wavelets, and in particular the Haar wavelet, provide a very simple approach to smoothing signals.

The 2×2 Haar matrix is given by

$$H_2 = \frac{1}{\sqrt{2}} \begin{bmatrix} 1 & 1 \\ 1 & -1 \end{bmatrix}.$$

Given an MC sample of $f(\cdot)$ whose length N is a power of 2 $(y_1, y_2, \ldots, y_{2^n-1}, y_{2^n})$, i.e. $N = 2^n$, and where the sample is sorted with respect to the input factor X_i under analysis, we may group its elements as $((y_1, y_2), \ldots, (y_{2^n-1}, y_{2^n}))$ and we may right-multiply each term by the matrix H_2,

$$H_2 \begin{pmatrix} y_{2j-1} \\ y_{2j} \end{pmatrix} = \begin{pmatrix} s_j \\ d_j \end{pmatrix} \quad \text{for } j = 1, \ldots, 2^{n-1}$$

obtaining two new sequences $(s_1, \ldots, s_{2^{n-1}})$ and $(d_1, \ldots, d_{2^{n-1}})$. The sequence s gives the sum between two consecutive points while the sequence d gives the difference. Since the original signal has a length equal to 2^n, one can recursively apply the same procedure to s-sequences up to n times. For each $\lambda = 1, \ldots, n$ we call the associated sequence s^λ the λth Haar approximation coefficients and d^λ the λth Haar detail coefficients.

For each $\lambda = 1, \ldots, n$ we may use s^λ to create an approximation $\hat{f}^\lambda$ of the original signal f. This is obtained first by rescaling each value of s^λ by $\sqrt{2^\lambda}$ and then by replicating each of them 2^λ times.

Example Consider a sample of length $2^4 = 16$ from a function $Y = f(X) = (X-0.5)^2 + \varepsilon$, where $X \sim U[0, 1]$ and ε is a white noise normally distributed $N(0, 0.03)$:

$$f = (0.21, 0.19, 0.14, 0.1, 0.05, 0.01, 0.03, -0.01, 0.01,$$
$$-0.03, 0.01, 0.06, 0.04, 0.1, 0.13, 0.19).$$

METAMODELLING AND THE HDMR

- First stage ($\lambda = 1$). From the first pair of y-points we compute the first element of s^1: $(0.21 + 0.19)/\sqrt{2} = 0.2828$. Repeating this with all pairs of y we get

$$s^1 = (0.2828, 0.1697, 0.0424, 0.0141, -0.0141, 0.0495,$$
$$0.0990, 0.2263).$$

Scaling the first term of s^1 by $\sqrt{2}$ we get 0.2, which is replicated 2^1 times to construct the first two elements of $\hat{f}^1$. This is repeated to get

$$\hat{f}^1 = (0.2, 0.2, 0.12, 0.12, 0.03, 0.03, 0.01, 0.01,$$
$$-0.01, -0.01, 0.035, 0.035, 0.07, 0.07, 0.16, 0.16).$$

- Second stage ($\lambda = 2$): Replicating the steps of the first stage to the s^1 sequence, we get

$$s^2 = (0.32, 0.04, 0.025, 0.23)$$

and

$$\hat{f}^2 = (0.16, 0.16, 0.16, 0.16, 0.02, 0.02, 0.02, 0.02,$$
$$0.013, 0.013, 0.013, 0.013, 0.12, 0.12, 0.12, 0.12)$$

- ... and so on.

If we compare the graph of the original signal f and its first- and second-level approximations $\hat{f}^1, \hat{f}^2$ we obtain the following picture:

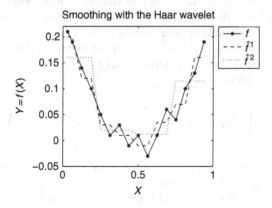

The picture above shows that $\hat{f}^\lambda$ may be used to infer the smoothed behaviour of the signal f, where λ plays the role of a smoothing parameter:

the higher λ, the smoother the signal extraction. At the highest level of smoothing $\lambda = n$, the Haar wavelet will simply provide the unconditional mean $E(Y)$. Note also that this smoothing approach can be implemented recursively, and does not require any matrix inversion.

Exercise Write a code for the Haar wavelet and reproduce the results shown in the previous example.

5.3.1.2 Spline smoothing

Smoothing splines are a nonparametric method which is very useful for understanding the more sophisticated methodology to be discussed later in this chapter. Given a Monte Carlo sample of size N, a cubic smoothing spline is a function $\hat{g}$ that minimizes the penalized residuals' sum of squares:

$$\sum_{j=1}^{N}[y_j - \hat{g}(x_j)]^2 + \lambda \int_a^b \left[\frac{d^2 \hat{g}(x)}{dx^2}\right]^2 dx \qquad (5.19)$$

where $a \leq \min(x_j)$, $b \geq \max(x_j)$, $j = 1, \ldots, N$ and λ is the Lagrange multiplier, which plays the role of a smoothing parameter (i.e. the bigger λ, the smoother $\hat{g}$). In (5.19), the first term is the sum of squared residuals, measuring fit to the data, while the second term penalizes too high a curvature in $\hat{g}$. This approach is also known in numerical analysis as *regularization* (or deterministic regularization, DR).

The unique, explicit solution to (5.19) is given by a natural cubic spline with knots at each observed value x_j (see Hastie and Tibshirani, 1990). A cubic polynomial spline is a function that is a cubic polynomial on any interval defined by adjacent knots, has two continuous derivatives and a third which is a step function that jumps at the knots.

Spline smoothing can easily be implemented by taking the discrete formulation of (5.19), where the long-term signal $\hat{t}$ is given by the solution of the minimization problem:

$$\min\left(\sum_{s=1}^{N}[y_s - \hat{t}_s]^2 + \lambda \sum_{s=2}^{N-1}[(\hat{t}_{s+1} - \hat{t}_s) - (\hat{t}_s - \hat{t}_{s-1})]^2\right), \qquad (5.20)$$

where the index s scans the data in a *sorted* order with respect to the input factor under analysis. In econometrics this is called the Hodrick–Prescott (HP) filter (Hodrick and Prescott, 1980), and it is used for trend extraction in economic time-series analysis (in such cases s scans the data in temporal order). As Young and co-workers have shown (Jakeman and Young, 1984; Young and Pedregal, 1999), this same problem (5.20) can

METAMODELLING AND THE HDMR

be solved by a very simple recursive filtering and smoothing algorithm that yields identical results to that of HP but is more sophisticated in statistical terms[2]. This recursive solution also has the advantage that it involves no matrix inversion, while the HP *en-bloc* solution, as we see below, involves the inversion of a matrix with dimension equal to that of the data length.

For the present, tutorial purposes, let us consider only the HP *en-bloc* solution, which can be formulated by taking derivatives with respect to $\hat{t}_s$: i.e.,

$$y_1 = (1+\lambda)\hat{t}_1 - 2\lambda\hat{t}_2 + \lambda\hat{t}_3$$
$$y_2 = -2\lambda\hat{t}_1 + (1+5\lambda)\hat{t}_2 - 4\lambda\hat{t}_3 + \lambda\hat{t}_4$$
$$\ldots$$
$$y_s = \lambda\hat{t}_{s-2} - 4\lambda\hat{t}_{s-1} + (1+6\lambda)\hat{t}_s - 4\lambda\hat{t}_{s+1} + \lambda\hat{t}_{s+2}$$
$$s = 3, \ldots, N-2$$
$$\ldots$$
$$y_{N-1} = \lambda\hat{t}_{N-3} - 4\lambda\hat{t}_{N-2} + (1+5\lambda)\hat{t}_{N-1} - 2\lambda\hat{t}_N$$
$$y_N = \lambda\hat{t}_{N-2} - 2\lambda\hat{t}_{N-1} + (1+\lambda)\hat{t}_N.$$

This can easily be put into matrix form,

$$\mathbf{y} = (\mathbf{I} + \lambda \mathbf{U}^T \mathbf{U}) \cdot \hat{\mathbf{t}}^\lambda, \tag{5.21}$$

where $\mathbf{I}$ is the $N \times N$ identity matrix, $\mathbf{y}$ is the $1 \times N$ vector of model output data and $\mathbf{U}$ is the following $N \times N$ matrix:

$$\mathbf{U} = \begin{bmatrix} 0 & 0 & 0 & 0 & \ldots & 0 & 0 & 0 \\ 0 & 0 & 0 & 0 & \ldots & 0 & 0 & 0 \\ 1 & -2 & 1 & 0 & \ldots & 0 & 0 & 0 \\ 0 & 1 & -2 & 1 & \ldots & 0 & 0 & 0 \\ \vdots & \vdots & \vdots & \vdots & \ddots & \vdots & \vdots & \vdots \\ 0 & 0 & 0 & 0 & \ldots & 1 & -2 & 1 \end{bmatrix}.$$

The explicit solution is easily found to be

$$\hat{\mathbf{t}}^\lambda = (\mathbf{I} + \lambda \mathbf{U}^T \mathbf{U})^{-1} \cdot \mathbf{y} = \mathbf{W}^{-1} \cdot \mathbf{y}. \tag{5.22}$$

---

[2] This IRWSM algorithm is available in the CAPTAIN Toolbox for Matlab, which can be downloaded from http://www.es.lancs.ac.uk/cres/captain/

The 'trend' $\hat{t}^\lambda$ is a nonparametric estimate of the univariate 'metamodel' $f_i(X_i) = E(Y|X_i)$, which approximates the function $f(\cdot)$. The dependence of such an estimate on the smoothing parameter is made explicit in (5.21) and (5.22) by the λ exponent in $\hat{t}^\lambda$. The appropriate value for λ is not immediately apparent: cross-validation, graphical methods and measures of degrees of freedom are used for this purpose (see Hastie and Tibshirani, 1990 and Storlie and Helton, 2008, for more details). Note that, as pointed out above, since the solution (5.21) involves the inversion of a $N \times N$ matrix, it can be computationally intensive and require a large memory size for large MC samples. Note also that, in the *en-bloc* approach, it is also possible to obtain the standard errors of the estimated t^λ. Readers can refer to Hastie and Tibshirani (1990) for a detailed discussion of *en-bloc* methods and estimation of standard error bands and to Young and Pedregal (1999) for a combined discussion of these topics in the case of recursive and *en-bloc* approaches.

Example Consider the same example used above for the Haar wavelet. Using the same data, we obtain the following estimates for $\hat{t}^\lambda$, at various smoothing levels:

- $\lambda = 1$:

  $\hat{t}^1 = (0.22, 0.18, 0.14, 0.096, 0.055, 0.026, 0.011,$
  $-0.0012, -0.0064, -0.0069, 0.01, 0.036, 0.059, 0.095, 0.14, 0.18).$

- $\lambda = 10$:

  $\hat{t}^{10} = (0.21, 0.17, 0.13, 0.095, 0.061, 0.033, 0.013,$
  $0.00069, -0.0035, 0.00068, 0.014, 0.036, 0.063, 0.096, 0.13, 0.17).$

- $\lambda = 100$:

  $\hat{t}^{100} = (0.17, 0.14, 0.12, 0.093, 0.071, 0.053, 0.039,$
  $0.03, 0.027, 0.029, 0.036, 0.048, 0.064, 0.082, 0.1, 0.12).$

Such smoothed estimates are shown in the plot at the top of next page.

This plot shows the smoothing effect of increasing λ. If λ tends to infinity, the HP-filter simply provides the unconditional mean $E(Y)$. Judging by eye, the best smoothing seems be obtained using λ values between 1 and 10.

Exercise Write a code for the spline smoothing (HP-filter) and reproduce the results shown in the previous example.

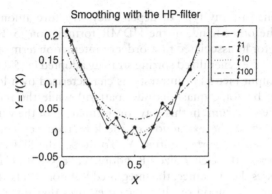

5.3.1.3 State-dependent regressions

Ratto et al. (2004, 2007) have recently presented a flexible and efficient approach to the estimation of $g_I^*(\mathbf{X}_I) = E(Y|\mathbf{X}_I)$ and of truncated ANOVA-HDMR decompositions. This estimation procedure is based on considering the HDMR as a State-Dependent Parameter (SDP) relationship (Young, 2000; Young et al., 2001) and, in particular, an example of the State-Dependent Regression (SDR) model estimation, as discussed in Young (2001). In this form, it is a non-parametric approach, based on recursive filtering and smoothing estimation procedures similar to those mentioned in the previous sub-section and available in the CAPTAIN Toolbox for Matlab (see earlier footnote).

In brief, the fundamental concept underlying the SDR approach is that any term like $E(Y|\mathbf{X}_I)$ can be viewed as an SDR model of the form

$$Y_s = p_{I,s}(\xi_I) + e_s \qquad (5.23)$$

where the state-dependent parameter (SDP) $p_{I,s}(\xi_I)$, $I = i_1, \ldots, i_l$, depends on a state variable ξ_I that moves, according to a generalized sorting strategy, along the coordinates of the single factors or groups of factors indexed by I; and e_s is the residual, i.e. the portion of variability of Y that cannot be explained by the group of factors indexed by I.

Extending this definition, the truncated ANOVA-HDMR expansion can also be expressed as an SDR model of the following form:

$$\begin{aligned} Y_s - f_0 &= \sum_i p_{i,s}(\xi_i) + \sum_{j>i} p_{ij,s}(\xi_{ij}) + \sum_{l>j>i} p_{ijl,s}(\xi_{ijl}) + e_s \\ &= \sum_i f_{i,s}(X_i) + \sum_{j>i} f_{ij,s}(X_i, X_j) + \sum_{l>j>i} f_{ijl,s}(X_i, X_j, X_l) + e_s \end{aligned} \qquad (5.24)$$

where e_s now represents the higher-order terms of the ANOVA-HDMR. Note also that, in the formulation (5.23), the SDP accounts for all cumulative effects within group I, be they first-order or interaction terms. As such, it can be applied to any type of dependency structure amongst the input factors. On the other hand, in the HDMR formulation (5.24), each SDP accounts only for its associated first-order or interaction term of the HDMR.

According to the generalized sorting strategy adopted in (5.23) and (5.24), the group of input factors of interest I is characterized by a low-frequency spectrum (e.g. by some quasi-periodic pattern) while the remaining ones present a white spectrum. In this way, the estimation of the various HDMR terms is reduced to the extraction of the low-frequency component (i.e. of a 'trend') from the sorted output Y. To do so, the SDPs are modelled by one member of the Generalized Random Walk (GRW) class of nonstationary processes. For instance, the Integrated Random Walk (IRW) process turns out to produce good results, since it ensures that the estimated SDR relationship has the smooth properties of a cubic spline[3].

Fixing ideas to the IRW characterization of each SDP, the model (5.24) can be put into the state-space form as

$$\text{Observation equation (HDMR):} \quad Y_s = \mathbf{z}_s^T \mathbf{p}_s + e_s$$
$$\text{State equations for each SDP:} \quad p_{I,s} = p_{I,s-1} + d_{I,s-1} \quad (5.25)$$
$$d_{I,s} = d_{I,s-1} + \eta_{I,s}$$

where $\mathbf{z}_s^T$ is the transposed regression vector, composed entirely of unity elements; $\mathbf{p}_s$ is the SDP vector; while e_s (observation noise) and $\eta_{I,s}$ (system disturbances) are zero-mean white-noise inputs with variance σ^2 and $\sigma^2_{\eta(I)}$ respectively. Given this formulation, the SDPs are estimated using the recursive Kalman filter and associated recursive Fixed Interval Smoothing (FIS) algorithm (Kalman, 1960; Bryson and Ho, 1969; Young, 1999b). The recursive state estimation requires that each SDP be estimated in turn, each with a different ordering of the data, within a backfitting procedure (Young, 2000, 2001). At each backfitting iteration, the hyperparameters associated to (5.25), namely the white noise variances σ^2 and $\sigma^2_{\eta(I)}$, are optimized by maximum likelihood (ML), using prediction error decomposition (Schweppe, 1965). In fact, by a simple reformulation of the KF and FIS algorithms, each SDP and its stochastic IRW process model can be entirely characterized by one Noise Variance Ratio (NVR) hyperparameter, where $\text{NVR}_I = \sigma^2_{\eta(I)}/\sigma^2$. Hence, only NVRs need to be optimized.

One very useful implication of the KF/FIS algorithms underlying SDR estimation is that the standard error $\sigma_{f,s}$ of the SDPs is recursively produced in a natural manner by reference to the covariance matrix of the state vector. This allows us to distinguish the significance of the estimated functions $E(Y|\mathbf{X}_I)$.

---

[3] Random Walk (RW) or Smoothed Random Walk (SRW) might be identified as being preferable in certain circumstances because they yield less smooth estimates.

The recursive formulation of the SDR model also implies a great flexibility in the estimations, whenever this is required, for instance in the case of heteroscedastic behaviour in the observation noise or in the case of discontinuities in the model output.[4] In practice, the basic SDR implementation used here can be easily extended to adapt to such situations, *typical in non-linear systems,* where one single, constant smoothing parameter λ does not allow us to follow appropriately the observed patterns of the $f(\cdot)$ mapping.

In its basic IRW formulation, the links between the SDR approach and the HP-filter smoothing are clear: both of them have the properties of a smoothing cubic spline. Moreover, it is also easy to verify the equivalence between the NVR and λ, linked by the simple relationship: $\lambda = 1/\text{NVR}$: see Young and Pedregal (1999). The NVR therefore plays the role of the inverse of a smoothing parameter. SDR advantages, however, are in terms of the ML estimation of the NVR, which makes the choice of the smoothing parameter completely objective, and in its great flexibility, as mentioned above. These properties provide optimal convergence properties of the SDR estimates to the best least-squares predictor of Y, given by $g^*(\mathbf{X}_I)$.

The SDR smoothing techniques based on GRW processes can also be seen as low-pass filters. In the case of IRW, the 50% cutoff frequency ω^* is linked to the NVR by the relationship $\text{NVR} = 4(1 - \cos(2\pi\omega^*))^2$ (see e.g. Young and Pedregal, 1999). The period $T^* = 2\pi/\omega^*$, which is obtained via the maximum likelihood estimation, can be compared to N to obtain an idea of 'typical' T^*/N ratios and to identify some rule-of-thumb criterion for the smoothing parameter λ of the HP-filter. Our experience suggests that a reasonable rule of thumb can be to set $T^* = N/3 \longrightarrow N/2$ and to derive λ accordingly. In the last step of the SDR analysis, the smoothed nonparametric 'curves' obtained from the SDR model estimation are parameterized by, for example, a linear wavelet functional approximation or a summation of RBFs, allowing us to build a full metamodel to replace the original one.

Example Consider again the simple example used before. Using the same data, we obtain $\text{NVR} = 0.59$ from ML estimation. This implies that the equivalent optimal λ for the HP-filter would be about 1.7, matching the values that 'looked' acceptable in the HP-filter analysis. The plots of the SDR estimate and the equivalent HP-filter estimate using $\lambda = 1/\text{NVR}$ are shown next:

---

[4] See Young and Ng (1989); Young and Pedregal (1996); Young (2002) and the discussion in Ratto *et al.* (2007).

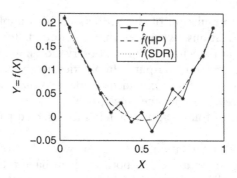

This clearly shows the identity between the HP-filter and IRW smoothing (the curves are the same), as well as the importance of the SDR estimation procedure linked to maximum likelihood, providing for optimal identification of the smoothing parameter rather than the *ad-hoc* tuning used in the HP-filter.

5.3.1.4 Estimating sensitivity indices

Once the smoothed estimates $\hat{g}_I$ of $E(Y|\mathbf{X}_I)$ have been obtained, the estimation of sensitivity indices $S_I = V[E(Y|\mathbf{X}_I)]/V(Y)$ is straightforward. As discussed in Doksum and Samarov (1995), three estimators can be applied for this purpose. The one we use is

$$\hat{S}_I = \frac{N^{-1}\sum_{s=1}^{N}(\hat{g}_I(\mathbf{x}_{I,s}) - \bar{g})^2}{\sigma_Y^2} \qquad (5.26)$$

where $\bar{g} = N^{-1}\sum \hat{g}_I(\mathbf{x}_{I,s})$ and $\sigma_Y^2 = N^{-1}\sum(y_s - \bar{Y})^2$. Doksum and Samarov (1995) also provide the error estimate for (5.26), by showing that $N^{1/2}(\hat{S}_I - S_I)$ is asymptotically normal with mean zero and variance $(1-S_I)^2 V[y^{*2} - u^2]$, where y^* and u are the standardized output and residual respectively, i.e. $y_s^* = (y_s - \bar{Y})/\sigma_Y$ and $u_s = [y_s - \hat{g}_I(\mathbf{x}_{I,s})]/(\sigma_Y(1-S_I)^{1/2})$. Hence, the standard error of the estimate of S_I is given by

$$SE(\hat{S}_I) = (1-\hat{S}_I)\text{std}[y_s^{*2} - u_s^2]/N^{1/2} \qquad (5.27)$$

5.3.2 A Simple Example

We give here an example of the smoothing estimation procedures. Let us consider the simple model

$$Y = X_1 + X_2^2 + X_1 \cdot X_2 \qquad (5.28)$$

with input distributions $X_i \sim N(0, 1)$. This model has the simple ANOVA-HDMR representation:

$$f_0 = 1$$
$$f_1(X_1) = X_1$$
$$f_2(X_2) = X_2^2 - 1$$
$$f_{1,2}(X_1, X_2) = X_1 \cdot X_2$$

We apply the smoothing procedures described above, by using an LP_τ sample of $N = 256$ model evaluations.

5.3.2.1 Haar wavelet smoothing

Let us first perform the smoothing with the Haar wavelet. As a rule-of-thumb criterion for the smoothing parameter of the Haar wavelet, we dictate that the extracted signal be made of $8 = 2^3$ values, i.e. we construct the smoothing by taking eight local averages from the sample of Y divided into eight bins. Given the sample of $N = 256 = 2^8 = 2^n$, this implies that $\lambda = 8 - 3 = 5$. The sample of the output Y has to be sorted according to each input factor and the Haar smoothing procedure described above has to be applied for each sorted sample.

The results of this procedure are shown in Figure 5.3. This shows that the rule of thumb of eight local means is able to provide an illustrative idea of the f_i patterns.

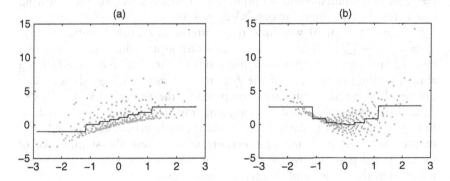

Figure 5.3 Haar estimation of the first-order HDMR of the simple model (5.28). (a, b) Scatterplots of Y versus X_1 and X_2 (grey dots), with the smoothed estimates of the $f_i + f_0$ functions (solid lines)

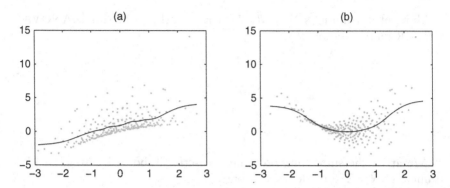

Figure 5.4 HP-filter estimation of the first-order HDMR of the simple model (5.28). (a, b) Scatterplots of Y versus X_1 and X_2 (grey dots), with the smoothed estimates of the $f_i + f_0$ functions (solid lines)

5.3.2.2 Spline smoothing (HP-filter)

In this case we have to choose the value for the smoothing parameter of the HP-filter. Using $\lambda = 50\,000$, which is in the range of rule-of-thumb values for the cut-off frequency ω^* described above, we get the results shown in Figure 5.4. We can observe the much nicer pattern provided by the cubic spline properties of the HP-filter with respect to the Haar wavelet. The problem, however, still lies in the approximate choice of λ.

5.3.2.3 SDR estimation

Here, the maximum likelihood optimization of the NVRs gives the following results for the two input factors: $NVR_1 = 4.46e\text{-}6$, $NVR_2 = 4.675e\text{-}5$.

In Figure 5.5 (a, b) we show the scatterplots of the modified output $Y_i^* = Y - f_0 - \sum_{j \neq i} f_j(X_j)$, used in the backfitting procedure, versus the two model parameters together with the result of the SDR recursive filtering and smoothing estimation of the f_i terms. In Figure 5.5 (c, d) we show the detail of the SDR estimates, compared to the analytical values (5.29). Apart from expectable border phenomena, the SDR estimates are excellent. In Figure 5.5 (c, d) the dashed lines show the 95% error bands ($= \pm 2 \cdot \sigma_{f,s}$) of the estimated patterns. This permits us to assess the significance of the estimated patterns by simply checking whether the zero-line is always included in the error band (implying insignificance) or not.

Naturally, given the equivalence of SDR and the HP-filter, using $1/NVR_i$ as smoothing parameters for the HP-filter analysis produces the same results as for the SDR analysis, provided that the backfitting procedure is also applied – otherwise the results will be slightly different.

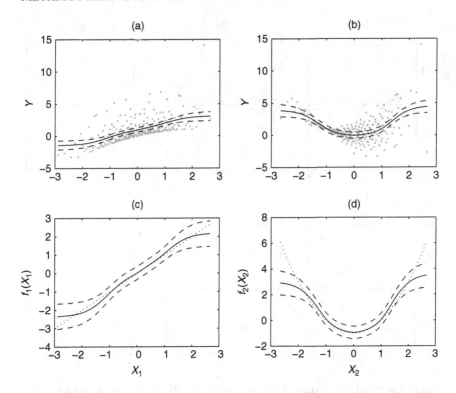

Figure 5.5 SDR estimation of the first-order HDMR of the simple model (5.28). (a, b) Scatterplots of the modified output $Y_i^* = Y - \sum_{j \neq i} f_j(X_j)$ used in the backfitting procedure versus X_1 and X_2 (grey dots), with the smoothed estimates of the $f_i + f_0$ functions (solid lines) and their 95% error bands (dashed lines). (c, d) Detail of the estimated f_i functions (solid lines) with estimated 95% error bands (dashed lines) and analytic f_i functions (dotted lines)

5.3.3 Another Simple Example

Let us consider the function

$$Y = X_1 \cdot X_2 + X_3 \tag{5.29}$$

with

$$X_i \sim U(-1, 1).$$

We perform an analysis using a Sobol' LP_τ sample of dimension 256. This model has only one non-zero main effect for X_3 and one second-order interaction term for (X_1, X_2). In Figure 5.6 we show the SDR estimation of the first-order HDMR terms f_1, f_2 and f_3. We also report the standard error band, showing that only f_3 has a significant main effect.

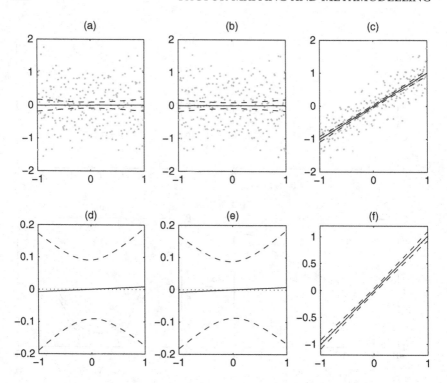

Figure 5.6 SDR estimation of the first-order HDMR of the simple model (5.29). (a, b, c) Scatterplots of Y versus X_i (grey dots), with the smoothed estimates of the $f_i + f_0$ functions (solid lines) and their 95% error bands (dashed lines). (d, e, f) Detail of the estimated f_i functions (solid lines) with estimated 95% error bands (dashed lines) and analytic f_i functions (dotted lines)

In order to give the flavour of the generalized sorting strategy used in the SDR methodology for interaction terms, let us perform the analysis of second-order interaction term (X_1, X_2). Since we want to compute the f_{12} interaction effect, the 2D sorting requires exploration of the (X_1, X_2) plane along a closed trajectory, like the one shown in Figure 5.7, with the sorting of the sample points carried out as they fall within the band delimited by two adjacent lines. This allows for the identification of an ordering in which (X_1, X_2) has low-frequency characteristics while X_3 maintains the white spectrum (Figure 5.8). The corresponding sorted output signal Y can then be analysed to identify the second-order interaction term. This is shown in Figure 5.9, where we compare the analytic values and the SDP estimates of the sorted $f_{1,2}$ interaction term.

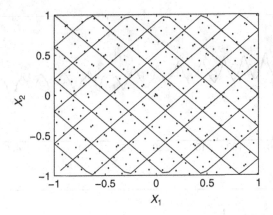

Figure 5.7 Sorting trajectory in the (X_1, X_2) plane

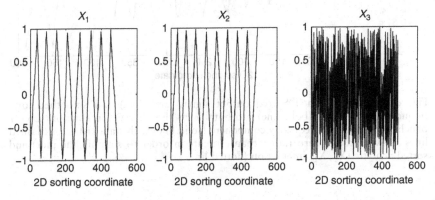

Figure 5.8 Sorted sample used to compute the (X_1, X_2) interaction

5.3.4 Exercises

1. Compute the analytic HDMR expansion of the function $Y = X_1 + X_2 + X_3$, with $X_i \sim N(\mu_i, \sigma_i)$. Assign values to (μ_i, σ_i) and perform the regression/smoothing analysis with the Haar wavelet and spline smoothing.
2. Compute the analytic HDMR expansion of the Ishigami function (Ishigami and Homma, 1990):

$$Y = \sin X_1 + A \sin^2 X_2 + B X_3^4 \sin X_1$$

where $X_i \sim U(-\pi, \pi)$. Perform the regression/smoothing analysis with the Haar wavelet and spline smoothing, when $A = 7$ and $B = 0.1$.

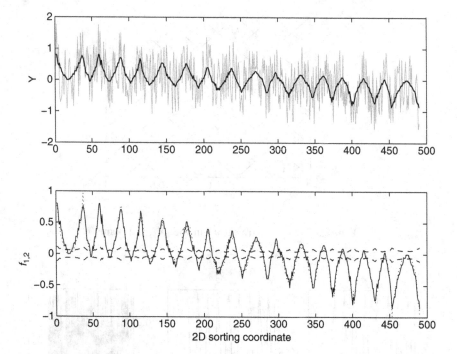

Figure 5.9 Upper panel: Sorted output signal (grey line) and smoothed low-frequency component (bold line) attributable to the (X_1, X_2) interaction. Lower panel: Smoothed low-frequency component (solid line) with standard error (dashed lines: shown around zero rather than estimate in order to simplify the plot) and sorted analytic (X_1, X_2) interaction (dotted line)

3. Compute the analytic HDMR expansion of the Sobol' g-function. Perform the regression/smoothing analysis with the Haar wavelet and spline smoothing, when:

    - $a = [0, 1, 4.5, 9, 99, 99, 99, 99]$;
    - $a = [0, 0, 1, 1, 4.5, 4.5, 9, 9, 99, 99, 99, 99, 99, 99, 99]$;
    - $a = [0, 0.01, 0, 0.2, 0.3, 0, 0.5, 1, 1.5, 1.8, 3, 4.5, 8, 9, 99]$;.

4. Compute the HDMR expansion of the model $Y = \prod_{i=1}^{k} X_i$, with $X_i \sim U(0, \text{MAX}_i)$. Is there a clever way to map this function synthetically?
    Perform the regression/smoothing analysis with the Haar wavelet and spline smoothing for the output Y and for the 'clever' transformation of Y, for $k = 3, 4$, assigning values to MAX_i.

5. Compute the HDMR expansion of the model $Y = \prod_{i=1}^{k} X_i$, with $X_i \sim N(0, \sigma_i)$. Is there a clever mapping in this case as well?

Perform the regression/smoothing analysis with the Haar wavelet and spline smoothing for the output Y and for the 'clever' transformation of Y, for $k = 3, 4$, assigning values to σ_i.

5.3.5 Solutions to Exercises

1. First we compute the unconditional mean

$$f_0 = E(Y) = \mu_1 + \mu_2 + \mu_3.$$

Then the first-order terms $f_i = E(Y|X_i) - f_0$:

$$f_1 = X_1 + \mu_2 + \mu_3 - f_0 = X_1 - \mu_1$$
$$f_2 = X_2 + \mu_1 + \mu_3 - f_0 = X_2 - \mu_2$$
$$f_3 = X_3 + \mu_1 + \mu_2 - f_0 = X_3 - \mu_3.$$

It is easy to verify that any other term of order higher than one is null.

2. Looking at the analytic form of the Ishigami function one can say that the decomposition will have the following terms:

$$Y = f_0 + f_1(X_1) + f_2(X_2) + f_3(X_3) + f_{13}(X_1, X_3).$$

Given the uniform probability density of X_i, we have that $p(X_i) = 1/2\pi$ if $-\pi < X_i < \pi$ and $p(X_i) = 0$ elsewhere. The unconditional mean is therefore

$$f_0 = E(Y) = \iiint_{-\pi}^{\pi} Y dX_1 dX_2 dX_3 / 8\pi^3$$
$$= \int_{-\pi}^{\pi} \sin X_1 dX_1 / 2\pi + \int_{-\pi}^{\pi} A \sin^2 X_2 dX_2 / 2\pi + \iint_{-\pi}^{\pi} BX_3^4 \sin X_1 dX_1 dX_3 / 4\pi^2$$
$$= 0 + A/2 + 0.$$

The first-order terms are

$$f_1 = E(Y|X_1) - f_0 = \iint_{-\pi}^{\pi} Y dX_2 dX_3 / 4\pi^2 - A/2$$
$$= \sin X_1 + \int_{-\pi}^{\pi} A \sin^2 X_2 dX_2 / 2\pi + \sin X_1 \int_{-\pi}^{\pi} B \cdot X_3^4 dx_3 / 2\pi - A/2$$
$$= \sin X_1 (1 + B \cdot 2\pi^5/5/(2\pi)) + A/2 - A/2$$
$$= \sin X_1 (1 + B\pi^4/5).$$

$$f_2 = E(Y|X_2) - f_0 = \iint_{-\pi}^{\pi} Y dX_1 dX_3 / 4\pi^2 - A/2$$

$$= \int_{-\pi}^{\pi} \sin X_1 dX_1/2\pi + A\sin^2 X_2 + \iint_{-\pi}^{\pi} BX_3^4 \sin X_1 dX_1 dX_3/4\pi^2 - A/2$$
$$= 0 + A\sin^2 X_2 + 0 - A/2$$
$$= A(\sin^2 X_2 - 1/2).$$
$$f_3 = E(Y|X_3) - f_0 = \iint_{-\pi}^{\pi} Y dX_1 dX_2/4\pi^2 - A/2$$
$$= (1 + BX_3^4) \cdot \int_{-\pi}^{\pi} \sin X_1 dX_1/2\pi + \int_{-\pi}^{\pi} A\sin^2 X_2 dX_2/2\pi - A/2$$
$$= 0 + A/2 - A/2 = 0.$$

It is easy to verify that all second-order terms are null except:

$$f_{13} = E(Y|X_1, X_3) - f_1 - f_3 - f_0$$
$$= \sin X_1 + BX_3^4 \sin X_1 + \int_{-\pi}^{\pi} A\sin^2 X_2 dX_2/2\pi$$
$$\qquad - \sin X_1(1 + B\pi^4/5) - 0 - A/2$$
$$= \sin X_1 - \sin X_1 + BX_3^4 \sin X_1 - B\pi^4/5 \sin X_1 + A/2 - A/2$$
$$= B(X_3^4 - \pi^4/5)\sin X_1.$$

3. In order to compute the HDMR expansion of the Sobol' g-function,

$$f = \prod_{i=1}^{k} g_i(X_i) = \prod_{i=1}^{k} \frac{|4X_i - 2| + a_i}{1 + a_i}$$

with $X_i \sim U(0,1)$, it is useful to recall the result from Chapter 4:

$$\int_0^1 g_i(X_i) dX_i = \int_0^1 \frac{|4X_i - 2| + a_i}{1 + a_i} dX_i = 1$$

valid for any X_i and any a_i.
So the unconditional mean is

$$f_0 = \prod_i \int_0^1 g_i(X_i) dX_i = 1.$$

The first-order terms are

$$f_i = E(Y|X_i) - f_0$$
$$= g_i(X_i) \prod_{j \neq i} \int_0^1 g_j(X_j) dX_j - f_0$$
$$= g_i(X_i) - 1.$$

The second-order terms are

$$f_{ij} = E(Y|X_i, X_j) - f_i - f_j - f_0$$
$$= g_i(X_i)g_j(X_j) \prod_{l \neq (i,j)} \int_0^1 g_l(X_l)dX_l - f_i - f_j - f_0$$
$$= g_i(X_i)g_j(X_j) - (g_i(X_i) - 1) - (g_j(X_j) - 1) - 1$$
$$= g_i(X_i)g_j(X_j) - g_i(X_i) - g_j(X_j) + 1.$$

All higher-order HDMR terms are obtained recursively, considering that, for any group of input factors indexed by $I = (i_1, \ldots, i_l)$:

$$E(Y|X_I) = \prod_{i \in I} g_i(X_i).$$

The Sobol' g-function has nonzero interaction effects of any order.
4. The unconditional mean is

$$f_0 = E(Y) = \prod_{i=1}^k \int_0^{MAX_i} X_i dX_i / MAX_i$$
$$= \prod_{i=1}^k MAX_i/2.$$

The first-order terms are

$$f_i = E(Y|X_i) - f_0$$
$$= X_i \prod_{j \neq i} \int_0^{MAX_j} X_j dX_j / MAX_j - f_0$$
$$= X_i \prod_{j \neq i} MAX_j/2 - f_0$$
$$= (X_i - MAX_i/2) \prod_{j \neq i} MAX_j/2.$$

The second-order terms are

$$f_{ij} = E(Y|X_i, X_j) - f_i - f_j - f_0$$
$$= X_i \cdot X_j \prod_{l \neq (i,j)} \int_0^{MAX_l} X_l dX_l / MAX_l - f_i - f_j - f_0$$
$$= X_i \cdot X_j \prod_{l \neq (i,j)} MAX_l/2 - f_i - f_j - f_0$$
$$= (X_i X_j - X_i MAX_j/2 - X_j MAX_i/2 - MAX_i \cdot MAX_j/4) \prod_{l \neq (i,j)} MAX_l/2.$$

All higher-order terms are obtained recursively, remembering that

$$E(Y|X_I) = \prod_{i \in I} X_i \prod_{j \notin I} \text{MAX}_j/2.$$

As with the *g*-function, this example also has nonnull interaction effects of any order.

A clever way to map the function Y is to consider its log-transformed version. This is allowed, since $X_i > 0$ for all X_i. This gives

$$\log(Y) = \sum_{i=1}^{k} \log(X_i)$$

which has an easy HDMR expansion of only first-order terms that describes 100% of Y!

In general, trying to analyse the log-transformed version of a function can produce interesting results. In particular, if the the log-transformation of Y is described up to a large extent (e.g. > 90% of the variance of $\log(Y)$) by first-order HDMR terms, this suggests that the mapping between X_i and Y can be factorized.

5. All the HDMR terms of this model are null except for the highest, kth order term:

$$f_{1,\ldots,k} = \prod_{i=1}^{k} X_i = f.$$

This is a very unattractive model. Even if the input factors had different σ_i values (i.e. different uncertainties), neither variance-based sensitivity indices nor HDMR analysis would be able to produce a ranking of the importance of input factors.

One possible way of analysing sensitivity for this kind of model is to apply MCF techniques and, similarly to Exercise 6 of Section 5.2.3, to try to map quantiles of Y, e.g. $Y > 1$ or $Y < -1$ (we leave this as an additional exercise for the reader).

In metamodelling terms, however, one can still identify a clever transformation of Y, extracting useful hints about the effect of each input factor on the outcome Y.

The conditional expectation $E(Y|X_I)$ is null whatever the single input factor or group of input factors. So, Y is a pure 'noise' process, without any shift in the mean, and the effect of each input factor is simply to modulate the noise amplitude, similarly to heteroscedastic processes.

One could therefore conceive of trying to build a metamodel for the variance of Y. In order to do this, one could consider the transformation $\log(Y^2)$. By doing so, the function to be analysed becomes

$$\log(Y^2) = \sum_{i=1}^{k} \log(X_i^2)$$

which is a simple, additive mapping between the (log of the) squared input factors and the (log of the) squared output. Remembering that any expectation of Y (unconditional or not) is null, taking the square of Y is equivalent to taking its variance. Then, the main effect $f_i = E\left(\log(Y^2)|\log(X_i^2)\right)$ describes effect of X_i in modulating the variance of Y, and the further X_i is from zero, the greater the variance of Y. So, the greatest impact in modulating the variance of Y will come from the input factor with the highest variance σ_i^2. This allows for a nice ranking of the importance of input factors, which will be proportional to the magnitude of the σ_i's.

This 'extreme' example shows that, in nasty cases where 'standard' global sensitivity analysis tools seem to fail in providing clear answers, functional transformations of Y can facilitate a better explanation of the model properties.

5.4 CONCLUSIONS

Chapter 4 demonstrated best available practices for computing sensitivity indices based on points thrown into a k-dimensional space. In this chapter we have discussed methods capable of greatly accelerating the computation of the indices (in terms of reduced number of model executions), based on metamodelling. All smoothing and metamodelling techniques are based on fitting a model approximation based on a single Monte Carlo (or, better, quasi-Monte Carlo) sample. This is a major advantage with respect to 'classical' variance-based techniques, which require some ad hoc sampling. Moreover, metamodelling techniques usually converge much more quickly, i.e. they are more efficient. This is possible since such techniques rely on assumptions of regularity and smoothness in Y that allow us to infer the value of Y at untried points, based on the information available from MC samples at nearby points. This also implies that, in contrast to 'classic' variance-based estimators which rely only on square-integrability of Y, smoothing methods are not robust in the face of heavily discontinuous mappings, e.g. piecewise continuous functions, in which the values of Y jump continuously in an apparently random fashion (i.e. presenting patterns similar to the one shown in Chapter 2, Figure 2.1(f)). In this regard, the recursive approach of Ratto *et al.* (2007) provides elements of meaningful flexibility to adjust the estimated model approximation to local jumps and spikes, provided that these are not as numerous as in Figure 2.1(f).

When to use what? The decision is evidently based on a trade-off between model execution cost and the analyst's time.

- The method of Saltelli (2002) is straightforward to encode and comes at the cost of $N(k+2)$ model executions, with $N = 1000$ or higher, to compute *both* the S_i's and the S_{Ti}'s.
- The method of Tarantola *et al.* (2006) described in Chapter 4 is also easy to encode and requires only a *single* set of N runs to compute the whole of set of S_i's. However, any information about interaction terms or total effects is missed.
- The approach of Ratto *et al.* (2004, 2007) discussed in this chapter is, according to our experience, an excellent metamodelling practice in terms of model executions, robustness of estimation and flexibility of implementation. It can give a fairly precise estimate of all indices of the first order at the overall cost of 200–500 simulations, although to obtain *reliable* estimates of second- and third-order interaction terms, somewhat longer samples are needed: e.g. 1000–2000. In its full formulation, it is less simple to encode with respect to the previous ones. Other methods mentioned here based on parametric regressions using polynomial bases to estimate HDMR terms could be helpful at a preliminary stage of identification of the most relevant terms for explaining the variability of Y.
- The total effect estimation is a weak element of metamodelling. Such an estimate requires adding all the first-order and interaction terms associated with each input factor. However, the precision of the mapping of the function $f(\cdot)$ is inversely proportional to the order of interactions, while the maximum order of interactions that can be reliably estimated with metamodelling techniques can hardly exceed the order of three, if one wants to reduce to the minimum the number of model evaluations. This implies that some relevant high-order interaction might always be missed in the metamodelling exercise. In this case, jointly performing an EE analysis, as discussed in Chapter 3, would allow for a cheap and quite comprehensive assessment in terms of low-order ANOVA-HDMR terms and EE, which replaces variance-based total effects.

Ideally a model-building environment should have software for both approaches at its disposal, unless the model is truly inexpensive to run.

Software developed for SDR modelling is available at http://sensitivity-analysis.jrc.cec.eu.int/.

6
Sensitivity Analysis: From Theory to Practice

SCOPE OF THE CHAPTER. EXAMPLES. A COMPOSITE INDICATOR. FINANCIAL OPTIONS MODELLING. A CHEMICAL REACTOR. A SIMPLE ANALYSIS. WHEN TO USE WHAT.

In this chapter a few examples are offered of how the different sensitivity analysis methods can be put to use. The aim is to offer suggestions rather than prescriptions, as we did in our discussion of possible 'settings' in Chapters 1 and 4. The setting up of a sensitivity analysis will in general depend upon:

- number of uncertain factors and computational cost of the model;
- characteristic of the output of interest (e.g. variance of the output, tails of its distribution, ...);
- scope of the analysis;
- nature and dispositions of the recipients of the analysis (owners of the problem, stakeholders, ...).

The last bullet is a reminder that the audience to which the analysis is addressed may also influence its set-up, for example in the degree of sophistication allowed in the presentation of the results.

Global Sensitivity Analysis. The Primer   A. Saltelli, M. Ratto, T. Andres, F. Campolongo, J. Cariboni, D. Gatelli, M. Saisana and S. Tarantola   © 2008 John Wiley & Sons, Ltd

6.1 EXAMPLE 1: A COMPOSITE INDICATOR

6.1.1 Setting the Problem

Imagine that an international organization or academic institution tries to capture in a single number, usually termed a composite indicator or index, the relative performance of several countries or regions in a multidimensional field. The Consumer Price Index, for example, considers the costs of various items purchased by a typical household each month. When a basket of about 60 goods and services is combined in the Consumer Price Index, it offers a more complete picture of the relative cost of living in different countries than would the price of bread or fuel alone.[1] A Handbook on Composite Indicators (Nardo et al., 2005) offers a review and guidelines for composite indicator development which we aim to adhere to in the present example.

However appealing the idea of summarizing complex phenomena into single numbers may be, the development of a composite indicator is not straightforward. It involves both theoretical and methodological assumptions which need to be assessed carefully to avoid producing results of dubious analytic rigour (Saisana et al., 2005). Furthermore, a composite indicator is likely to be received by a polarized audience, a fact which calls for stringent standards of rigour and robustness (Saltelli, 2006).

Composite indicators tend to sit between advocacy (when they are used to draw attention to an issue) and analysis (when they are used to capture complex multidimensional phenomena), and for this reason both their production and their use in the policy discourse are on the increase. This is also due to the media hunger for the apparently simple 'facts' purveyed by these measures.

To maximize their utility and minimize their misuse, therefore, developers must base these indices on the best available evidence, document clearly their structures, and validate them using appropriate uncertainty and sensitivity analyses. In the example that follows, we address two main questions that a developer of a composite indicator may be asked to respond to regarding the reliability of the results: Given the uncertainties during the development of an index, which of them are the most influential in determining the variance of:

- the difference between the composite scores of two countries with similar performance? This question could be used to address stakeholders' concern about a suspected bias in the score of a given country.

---

[1] An extensive list of such indices from various fields, such as the economy, environment, society and globalization, is presented on an information server on composite indicators http://composite-indicators.jrc.ec.europa.eu/

- the countries' ranking? This question pertains to the overall 'plausibility' or robustness of the index.

This example will show the potential of using the Elementary Effects (EE) method and variance-based methods in tandem for conducting sensitivity analysis of nonlinear models that have numerous uncertain factors. In brief, if the model is both expensive to run and/or has numerous input factors, one can use the EE method to reduce the number of factors first, and can then run a variance-based analysis on a reduced set of factors. A Monte Carlo filtering application complements the case study. For didactic purposes, we have chosen to work with an example that can be easily reproduced and with negligible model execution time.

6.1.2 A Composite Indicator Measuring Countries' Performance in Environmental Sustainability

In general, an index is a function of underlying indicators. Weights are assigned to each indicator to express the relevance of the indicators in the context of the phenomenon to be measured. Assume that there are m countries, whose composite score is to be constructed based on q indicators. Let X_{ij} and w_i respectively denote the value for a country j with respect to indicator i and the weight assigned to indicator i. All indicators have been converted into benefit-type ones, so that higher values indicate better performance. The problem is to aggregate $X_{ij} (i = 1, 2, \ldots, q, j = 1, 2, \ldots, m)$ into a composite index $CI_j (j = 1, 2, \ldots, m)$ for each country j. The most commonly used aggregation method is the weighted arithmetic mean:

$$CI_j = \sum_{i=1}^{q} w_i X_{ij}. \qquad (6.1)$$

The popularity of this type of aggregation is due to its transparency and ease of interpretation. An alternative aggregation method is the weighted geometric mean:

$$CI_j = \prod_{i=1}^{q} X_i^{w_i}. \qquad (6.2)$$

One argument in favour of a geometric approach instead of arithmetic could be the feature of compensability among indicators. A sometimes undesirable feature of additive aggregation (e.g. weighted arithmetic mean) is the full compensability they imply, whereby poor performance in some indicators can be compensated by sufficiently high values of other indicators. To give an example, assume that a composite indicator is formed by four indicators: inequality, environmental degradation, GDP per capita and unemployment. Two countries, one with values [21, 1, 1, 1] and the other with [6, 6, 6, 6], would have equal arithmetic mean (= 6), but different geometric mean (2.14 and 6, respectively). In this case, the geometric approach properly reflects the different social conditions in the two countries, which are, however, masked by the arithmetic aggregation.[2]

We will base our example on the Pilot 2006 Environmental Performance Index developed by the Yale and Columbia universities and described in Esty *et al.* (2006). The index focuses on a core set of environmental outcomes linked to 16 policy-related indicators for which governments in more than 130 nations should be held accountable. However, for didactic purposes and with a view to provide an easy to reproduce case study, we will restrict ourselves to a set of five indicators and six countries. The full, real-life example of sensitivity analysis for the Environmental Performance Index can be found in Esty *et al.* (2006). Table 6.1 lists the indicators that are included in our simplified composite indicator. There are two basic indicators related to air quality (regional ozone concentrations and urban particulate matter concentrations), two indicators related to sustainable energy (energy efficiency and CO_2 emissions per GDP) and finally the percentage of the population with access to drinking water.

The values for all five indicators and six countries are assumed to be uncertain in this exercise. For example, it could be that the set of countries we are studying was lacking these indicator values, which were therefore estimated by means of multiple imputation (Little and Rubin, 2002), or the data were provided by a statistical agency with an estimate of their sampling error. Data imputation could help to minimize bias and the need for 'expensive-to-collect' data that would otherwise be avoided only by omitting the relevant countries from the analysis. However, imputation may also influence the index. The uncertainty in the imputed data is reflected

Table 6.1 List of indicators in the simplified Environmental Performance Index

Abbreviation	Description
OZONE	Regional ozone
PM10	Urban particulate matter
ENEFF	Energy efficiency
CO2GDP	CO_2 emissions per GDP
WATSUP	Drinking water

---

[2] For a proper discussion of compensability and other theoretical aspect of composite indicators building see Nardo *et al.* (2005).

EXAMPLE 1: A COMPOSITE INDICATOR

Table 6.2 Distributions (normal: μ,σ) for the inputs of the simplified index and the trigger to decide the aggregation method (discrete distribution)

Country	OZONE	PM10	ENEFF	CO2GDP	WATSUP
A	(25.3, 10.1)	(53.3, 6.4)	(78.5, 3.9)	(56.3, 7.7)	(76.5, 10.5)
B	(29.0, 10.0)	(37.4, 6.6)	(58.9, 4.3)	(58.8, 7.6)	(56.7, 9.9)
C	(35.0, 9.9)	(26.8, 6.7)	(69.2, 4.1)	(43.8, 7.9)	(20.6, 8.9)
D	(78.7, 9.2)	(33.2, 6.6)	(77.9, 3.9)	(73.3, 7.4)	(27.8, 9.1)
E	(24.1, 10.1)	(32.2, 6.6)	(61.3, 4.3)	(66.9, 7.5)	(62.1, 10.1)
F	(30.5, 10.0)	(54.6, 6.4)	(67.3, 4.1)	(37.2, 8.1)	(22.4, 8.9)
Weight	(0.21, 0.05)	(0.25, 0.05)	(0.15, 0.05)	(0.23, 0.05)	(0.16, 0.05)

Aggregation: Discrete. 0: linear, 1: geometric, selected with equal probability

by variance estimates. This allows us to take into account the impact of imputation in the course of the analysis. The weights that are attached to these indicators are also considered uncertain for the sake of the example. Weights could derive either from statistical approaches (e.g. factor analysis, data envelopment analysis) or by consulting experts. In any case, different approaches will result in different values for the weights. In our example, all indicators values and weights follow a normal distribution. Mean and standard deviation values are reported in Table 6.2. Additionally, a trigger decides on the aggregation method for the basic indicators. The trigger is sampled from a discrete distribution, where 0: linear aggregation, 1: geometric aggregation.

6.1.3 Selecting the Sensitivity Analysis Method

In a general case, the validity and robustness of a composite indicator could depend on a number of factors, including:

- the model chosen for estimating the measurement error in the data;
- the mechanism for including or excluding indicators in the index;
- the indicators' preliminary treatment (e.g. cutting the tails of a skewed distribution, or removing outliers);
- the type of normalization scheme applied to the indicators to remove scale effects;
- the amount of missing data and the imputation algorithm;
- the choice of weights attached to the indicators;
- the choice of the aggregation method.

All these assumptions can heavily influence countries' scores and ranks in a composite indicator, and should be taken into account before attempting an interpretation of the results. In this example, we will examine three of the sources of uncertainty listed above: the indicators' values, the weights, and the aggregation method (weighted arithmetic mean versus weighted geometric mean). The questions we will address are:

- which are the most influential input factors in the difference between the composite scores of two countries with similar performance?
- which are the most influential input factors in the countries' ranking?

The latter question can be more closely focused on our problem by asking: if we could invite experts to reach consensus on some of the weights, which ones should they focus on with a view to reducing the output variance the most? Which indicators' values need to be re-examined given that they affect the output variance the most?

6.1.4 The Sensitivity Analysis Experiment and Results

The baseline scenario for the composite indicator model is the weighted arithmetic mean, where all indicators' values and weights are at their nominal value. In this scenario (see Table 6.3), country D would rank on top with a score of 57.8 $(= 78.7 \times 0.21 + 33.2 \times 0.25 + 77.9 \times 0.15 + 73.3 \times 0.23 + 27.8 \times 0.16)$, while country C has the lowest performance of the six countries (score of 37.8).

When the three sources of uncertainty are acknowledged, then each country score is no longer a single number, but a random variable. In total, there are 36 input factors in our sensitivity analysis (56 indicators values, 5 weights, 1 trigger to decide the aggregation method).

To get an idea of how the distributions of the countries' scores look when uncertainties are acknowledged, we present in Figure 6.1 the relevant histograms produced by 10 000 random combinations of the input factors. The countries' scores look roughly Gaussian, and the standard deviations among them are very similar, ranging from 4.4 for country B to 5.5 for country A.

We can easily observe that the histograms for several countries display significant overlap, e.g. countries A and D, B and E, and C and F. In

Table 6.3 Composite indicator scores for six countries, baseline scenario (weighted arithmetic mean). Numbers in parentheses represent ranks

A	B	C	D	E	F
55.6 (2)	46.9 (4)	37.8 (6)	57.8 (1)	47.6 (3)	42.3 (5)

EXAMPLE 1: A COMPOSITE INDICATOR

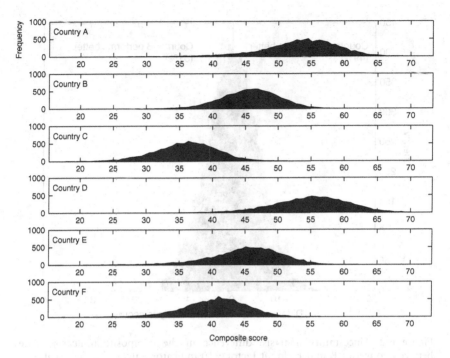

Figure 6.1 Uncertainty analysis: Composite indicator scores for six countries when accounting for uncertainties in the indicators' values, weights, aggregation method

all those cases, the message about the countries' relative performance, or the reasons for a good or bad performance, are not explicit. At the same time, the country rankings produced by the Monte Carlo simulations are different from the baseline ranking. This casts doubt on the belief that the composite indicator gives a fair representation of all scenarios regarding the countries' environmental performance. Yet, we could already conclude that the performance of the countries is of the order of: $(A, D) > (B, E) > (C, F)$. We will next study both issues, i.e. the reasons behind the overlap of two countries' scores, and the cumulative shift from the baseline ranking.

Figure 6.2 presents the histogram of the differences between the composite scores of countries B and E, which correspond to 10 000 Monte Carlo runs. Country B performs better than country E in 51% of the cases. It would be interesting to find out which uncertainties are driving this result. Since we are looking for important factors and we do not know a priori if the model is additive or not, we will use a variance-based method because of its model independence (appropriate for nonlinear and nonadditive models). However, from the exercises in Chapter 1, the reader will suspect that geometric aggregation will result in a nonadditive model.

The 16 input factors (2 countries ×5 indicator values, plus 5 weight values, and 1 trigger to decide the type of aggregation) are sampled using a

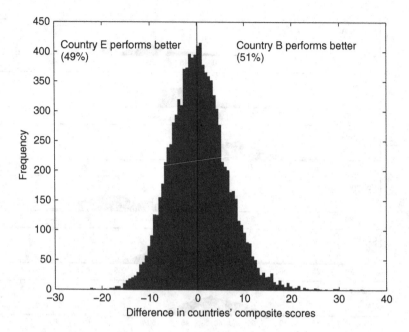

Figure 6.2 Uncertainty analysis: difference in the composite indicator scores between countries B and E. Input factors: 10 indicator values, 5 weight values, 1 trigger (weighted arithmetic mean versus weighted geometric mean)

quasi-random scheme (Sobol', 1967) of size $N = 1024$ and the composite scores per country are calculated performing 18 432 simulations (recall that based on the computational recipes in Chapter 4, for a model of k factors, only $N(k+2)$ model runs are needed). The sensitivity indices S_i and S_{T_i} are calculated for the score of countries B and E, and for their difference, and reported in Table 6.4. The first-order sensitivity indices, S_i, show that the interactions between input factors do not have an impact on the score for either country B or E, since the output variance explained by the single factors is 92.5% for country B and 90.6% for country E. The composite score for country B is mainly influenced by the uncertainty in the values of OZONE, PM10 and CO2GDP. Similarly, the variance of the composite score for country E is mostly attributed to OZONE and PM10 and the choice of the aggregation method. None of the weights has a significant impact on the composite scores of the two countries.[3] Regarding the difference between the two countries' composite scores, this is mostly attributed to the values for OZONE for both countries. It is interesting to

---

[3] In a real-life application the noninfluence of weights could be of crucial importance in the discussion among stakeholders and could facilitate acceptance of the index.

EXAMPLE 1: A COMPOSITE INDICATOR

Table 6.4 Sensitivity indices for the composite scores of two countries (B and E) and their scores' difference

Input factors	$S_{i,Bscore}$	$S_{i,Escore}$	$S_{i,(B-E)score}$	$S_{Ti,(B-E)score}$
OZONE(B)	0.369	-	0.192	0.210
PM10(B)	0.155	—	0.079	0.079
ENEFF(B)	0.013	-	0.007	0.009
CO2GDP(B)	0.112	-	0.059	0.061
WATSUP(B)	0.090	-	0.047	0.053
OZONE(E)	-	0.343	0.274	0.332
PM10(E)	-	0.117	0.087	0.091
ENEFF(E)	-	0.009	0.007	0.008
CO2GDP(E)	-	0.062	0.046	0.051
WATSUP(E)	-	0.049	0.036	0.042
w(OZONE)	0.043	0.051	0.000	0.017
w(PM10)	0.011	0.017	0.004	0.005
w(ENEFF)	0.017	0.019	0.000	0.001
w(CO2GDP)	0.019	0.026	0.002	0.005
w(WATSUP)	0.010	0.019	0.001	0.006
Aggregation	0.086	0.194	0.023	0.096
Sum	0.925	0.906	0.864	1.064

note that PM10, despite its influence on both countries' scores, does not have an impact on their difference. The 16 input factors taken individually account for 86.4% of the variance of the difference of the two countries scores, while the remaining 13.6% is due to interactions.

The next issue we come to is: which factors have the strongest impact on differences in the country ranking with respect to the baseline ranking? To express this mathematically, we calculate the cumulative shift from the baseline ranking, so as to capture in a single number the relative shift in the position of the entire system of countries. This shift can be quantified as the sum over m countries of the absolute differences in countries' ranks with respect to the baseline ranking:

$$R_S = \sum_{j=1}^{m} \left| Rank_{baseline}(CI_j) - Rank(CI_j) \right| \tag{6.3}$$

Figure 6.3 provides the histogram of 10 000 random Monte Carlo calculations of the cumulative shift from the baseline ranking. In most cases the shift with respect to the baseline ranking is of 2, 4 or 6 positions in total.

Our aim is to identify which of the 36 input factors (5 × 6 indicators' values, 5 weights, 1 trigger) are highly influential in our output, namely the cumulative shift in countries' ranks. Although, the computational cost of running the model is not high for this simple model we will assume that

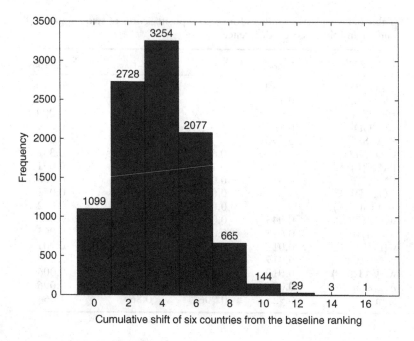

Figure 6.3 Uncertainty analysis. Output: cumulative shift from the baseline ranking for six countries. Uncertain input factors: 30 indicator values, 5 weight values, 1 trigger to decide the aggregation method (arithmetic vs geometric mean)

it is to illustrate an application of factor screening. We therefore resort to a screening method to help us reduce the number of input factors. We use the EE method described in Chapter 3 employing the recommended configuration of four levels for each input factor and ten trajectories. The total number of model executions performed to estimate the *EET* is 370 (=number of trajectories × (number of factors +1)).

Figure 6.4 shows the EET value for the 36 input factors. The most influential factors are the aggregation scheme, the weights for OZONE, WATSUP and PM10, and the indicators' values of OZONE(E), CO2GDP(B) and WATSUP(F). At the lower end, 13 factors are screened as nonimportant (*EET* < 0.5). By fixing these noninfluential input factors to their nominal values, we can study the impact of the remaining 23 factors more thoroughly using a variance-based approach. Recall that our aim is to identify the indicator values and weights that need to be re-estimated, so that the country ranking is not strongly affected by the remaining uncertainties. The total number of model executions to estimate the first-order and total effect sensitivity indices $\{S_i, S_{T_i}, i = 1, \ldots, 23\}$ is 25 600 (1024 × (23 + 2) where 1024 is the base sample), using the recipe in Chapter 4. Results are reported in Table 6.5. The impact on the cumulative shift in countries' ranks on the

EXAMPLE 1: A COMPOSITE INDICATOR

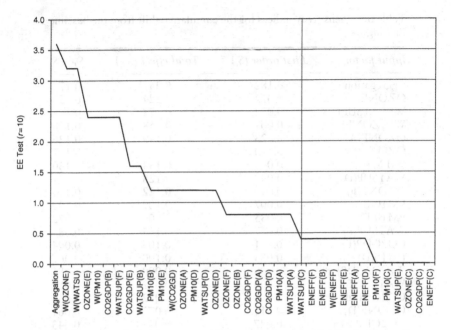

Figure 6.4 Results of the Elementary Effect screening exercise for 36 input factors

input factors is highly nonlinear. Only 56% of the output variance can be explained by the single factors, a great part of which is attributable to the aggregation method and the value of OZONE for country E. The aggregation method is the most influential both in single terms (highest S_i) and through interactions (highest S_{T_i}). The greatest S_{T_i}'s among the indicator values are found for OZONE in most countries, followed by PM10 for country E and WATSUP for country D. It is interesting to note that the estimated value for PM10 for country E has a relatively low standard deviation compared to other indicators' values (for example those of CO2GDP), yet the analysis showed that this factor is very influential due to interactions. Regarding the weights, although they have low values, they become very influential due to interactions – in particular w(OZONE), w(PM10) and w(WATSUP). Overall, the interactions and the nonlinearity account for 44% (= 100 − 56%) of the variance in the cumulative shift in countries' ranks. This result could guide the developers of the composite indicator to optimal resource allocation in terms of time and money spent to improve the reliability of just the few indicators' values that are most crucial to the policy message conveyed by the composite score or rank.

As a final consideration in our example, we will assume that a developer of a composite indicator is interested in a particular form of mapping. The objective of the analysis is to measure what fraction of the model realizations

Table 6.5 Sensitivity indices for the cumulative shift from the baseline ranking

Input factor	First order (S_i)	Total effect (S_{T_i})	$S_{T_i} - S_i$
Aggregation	0.182	0.357	0.175
OZONE(E)	0.072	0.224	0.152
W(WATSUP)	0.022	0.178	0.157
W(OZONE)	0.001	0.158	0.158
OZONE(A)	0.023	0.149	0.126
OZONE(B)	0.020	0.147	0.127
W(PM10)	0.004	0.134	0.130
WATSUP(D)	0.038	0.128	0.091
OZONE(F)	0.019	0.122	0.104
PM10(B)	0.007	0.109	0.103
PM10(E)	0.035	0.106	0.072
WATSUP(F)	0.017	0.105	0.088
CO2GDP(F)	0.011	0.105	0.094
PM10(D)	0.033	0.100	0.067
W(CO2GDP)	0.008	0.090	0.081
WATSUP(B)	0.000	0.082	0.082
CO2GDP(B)	0.014	0.082	0.068
OZONE(D)	0.011	0.072	0.061
CO2GDP(E)	0.027	0.070	0.043
PM10(A)	0.001	0.066	0.065
CO2GDP(D)	0.008	0.060	0.053
WATSUP(A)	0.003	0.053	0.050
Sum	0.558	2.745	2.187

fall within established bounds or regions. This objective can be pursued using Monte Carlo filtering (MCF) that we discussed in detail in Chapter 5. Recall that in MCF one samples the space of the input factors as in ordinary MC, and then categorizes the corresponding model output as either within or without the target region (the terms behaviour, B, or nonbehaviour, $\bar{B}$, are used). This categorization is then mapped back onto the input factors, each of which is thus also partitioned into a behavioural and nonbehavioural subset. When the two subsets B, $\bar{B}$ for a given factor are proven to be statistically different (by employing a Smirnov test for example), then the factor is an influential one. In our example we set the model's target behaviour as $R_s \leq 2$ (see Equation 6.3), that is we accept as behavioural an overall shift in rank of a maximum of two positions. Then, an MCF procedure can identify a significant correlation (> 0.2) between three pairs of input factors, OZONE(B)–WATSUP(B), OZONE(E)–CO2GDP(E), OZONE(F)–CO2GDP(F). The corresponding patterns are shown in Figure 6.5. The correlation coefficients are negative and slightly over 0.2. This implies that in order to have practically no shift in the countries' ranks with respect to the baseline scenario, then high values of OZONE have to be combined with

EXAMPLE 1: A COMPOSITE INDICATOR

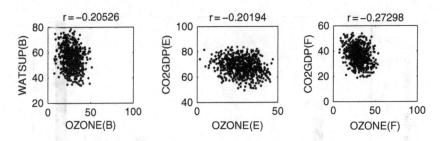

Figure 6.5 Bidimensional projection of the B subset for the three pairs of input factors that have significant correlation coefficients (> 0.2 in the MC sample). Target behaviour $R_s \leq 2$ (see Equation (6.3))

low values of WATSUP for country B, and vice versa. The same high–low combination of values should be observed for country E and country F, but in these cases the variables that play a role are OZONE and CO2GDP.

In order to gain further insight into the combinations of input factors that lead to high values of the cumulative shift in ranking (e.g. $R_s > 8$ in the $\bar{B}$ subset), we apply a similar approach. The MCF identifies the pair OZONE(E)–WATSUP(D) as the one having the highest correlation coefficient (−0.24) in the $\bar{B}$ subset. Figure 6.6 presents the bidimensional

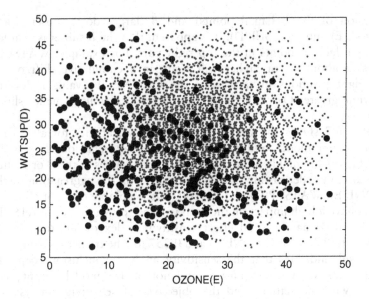

Figure 6.6 Bidimensional projection of the B (small points) and $\bar{B}$ (larger dots) subsets for OZONE(E) and WATSUP(D). This pair of input factors had the highest correlation (−0.24) in the $\bar{B}$ subset. Target behaviour $R_s \leq 8$

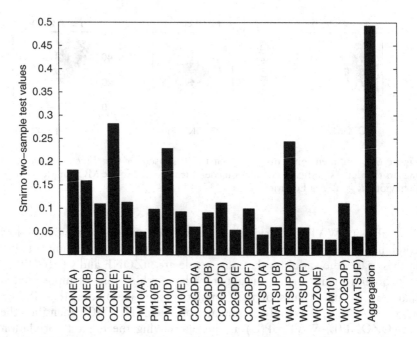

Figure 6.7 Smirnov two-sample test values (two-sided version) for all 23 input factors in the model. Target behaviour $R_s \leq 8$

projection of the B (small points) and $\bar{B}$ (larger dots) subsets for the OZONE(E) and WATSUP(D). It is apparent that the $\bar{B}$ subset is characterized by a lack of points in the top right part. For example, WATSUP(D) values at about 45 or higher will result in a cumulative shift of more than eight positions in ranking if the OZONE (E) value is less than 15, while greater OZONE(E) values will result in much smaller shifts in ranking.

Figure 6.7 shows the values of the Smirnov statistics for all 23 input factors in our model, and considering the target behaviour $R_s \leq 8$. Four input factors are highly influential in the target values in the B or $\bar{B}$ subset. These are, in order of importance, the trigger on the aggregation method, OZONE(E), WATSUP(D) and PM10(D). It is notable that not all of these factors have a high impact on the total variance of R_s. In fact, WATSUP(D) and PM10(D) have a total effect sensitivity index of less than 0.13, as opposed to 0.224 for OZONE(E) and 0.357 for the trigger on aggregation. This result indicates that the sensitivity analysis conclusions depend on the question we want to answer and not on the model output per se. This is why the settings and the objectives of sensitivity analysis must be clearly stated at the outset. Figure 6.8 is a graphical representation of the Smirnov test for the most important parameters that decide model realizations $R_s > 8$ (dotted curves). Only parameters with Smirnov statistics

EXAMPLE 1: A COMPOSITE INDICATOR

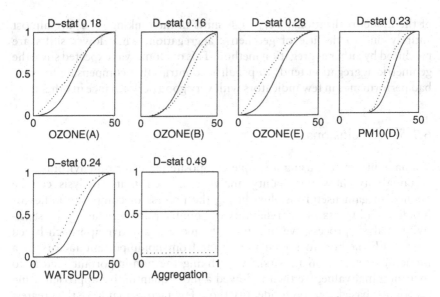

Figure 6.8 Smirnov test for the most important parameters (with Smirnov statistic > 0.15) that drive the partition into B ($R_s \leq 8$, continuous lines) and $\bar{B}$ (dotted lines) model realizations

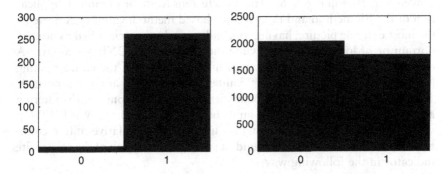

Figure 6.9 Histograms of the $\bar{B}$ subset ($R_s > 8$, left graph) and B ($R_s \leq 8$, right graph) for the trigger

greater than 0.15 are shown. Model realizations in the $R_s > 8$ subset are more likely when OZONE(A), OZONE(E), PM10(D) and WATSUP(D) fall on the left part of their predefined range, or when OZONE(B) falls on the right part of its predefined range. It is also interesting to note that the cumulative distribution for aggregation in the $R_s > 8$ subset is almost entirely concentrated in 1, i.e. for geometric aggregation.

The impact of the aggregation trigger is better depicted in Figure 6.9. The figure evidences the high impact of the aggregation trigger on the partitioning of the model realizations. Very high shifts in the ranking (> 8) are almost entirely due to the use of geometric aggregation, while lower shifts are produced by either aggregation method. This outcome was expected since the geometric aggregation tends to penalize countries that compensate for very bad performance in few indicators with very good performance in others.

6.1.5 Conclusions

We have illustrated, using a simple composite indicator on environmental sustainability, how uncertainty analysis and sensitivity analysis can be applied to gain useful insights during the process of composite indicator building and to assess the reliability of country ranking. During the sensitivity analysis process, we first used the more qualitative approach based on the EE method to screen important from nonimportant factors in a numerous set (i.e. 36 factors). After fixing the non-important factors to their nominal value, we then followed a more quantitative approach using a variance-based method to identify those few factors that have the greatest impact on the variance of the cumulative shift from the baseline ranking. Finally, we applied Monte Carlo filtering to identify which input factors result in very high values of the cumulative shift from the baseline ranking.

We summarize in Figure 6.10 the conclusions from the in-tandem application of the EE method and the variance-based method. Country A presents the most extreme picture, having four indicators' values classified as nonimportant or of low importance, while the value for OZONE for country A is considered to be highly influential in the variance of the entire ranking (all six countries). All indicators' values for country C are nonimportant. The pattern for the remaining countries varies. The Monte Carlo filtering approach identified a few more input factors, PM10(D) and WATSUP(D), whose uncertainty can result in high values for the cumulative shift in countries' ranks. These results could guide a subsequent revision of the composite indicator in the following way:

- If the experts who assigned weights to the indicators can be engaged in a Delphi process to reach consensus on a few of the weights, then this should be done for the weights assigned to OZONE and WATSUP.
- If the amount of resources (in time/money) is sufficient for improving the quality of just a few indicators' values, then those should be the values of OZONE for countries A, B and E and of the PM10 and WATSUP for country D.

Of course even more stable results would be obtained if the experts could reach consensus on the aggregation method to be used, rather than using a combination of linear and geometric as in the present example.

EXAMPLE 2: IMPORTANCE OF JUMPS IN PRICING OPTIONS

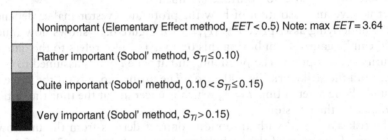

Figure 6.10 Classification of the 36 input factors regarding their impact on the cumulative shift from the baseline ranking. Methods used: first EE method, then variance-based

We have applied almost all methods described in this book to the composite indicator example. Clearly this is not to imply that all methods should always be used in combination on any given model. To compare abstraction with real life, the reader can study the sensitivity analysis performed on the actual Environmental Performance Index in Esty *et al.* (2006).

6.2 EXAMPLE 2: IMPORTANCE OF JUMPS IN PRICING OPTIONS

6.2.1 Setting the Problem

Imagine that an investor enters into a European call option contract that gives the holder the right (not the obligation) to buy an asset at a certain date T (e.g. two months) for a certain price. The price in the contract is known as the exercise price or strike price K; the date in the contract is known as the expiration date or maturity. To enter into an option contract there is a cost corresponding to the right purchased. This cost, referred to as premium $C(K, T)$, is the price of the option at present day and it is established according to the theory of arbitrage-free pricing (Hull, 1997).

Classic deterministic arbitrage involves buying an asset at a low price in one market and immediately selling it at a higher price in another market to make a risk-free profit. Instead, in the present exercise we will assume a market where arbitrage is not allowed, i.e. an arbitrage-free market or risk-neutral world. The theory of arbitrage-free prices stipulates that the prices of different instruments be related to one another in such a way that they offer no arbitrage opportunities. In practice, to price the option we make use of a model describing the evolution in time of the underlying asset price and then impose no arbitrage arguments.

Let us give an illustration of how the profit an investor makes depends on the underlying asset price $P = \{P_t, t \geq 0\}$. Figure 6.11 shows how much profit can be gained from buying this option. The plot refers to the date of maturity of the option. The premium paid at time $t = 0$ is assumed to be 5 euros, and the strike price is 100 euros. The profit depends on the price of the underlying asset at time $t = T$, which is uncertain at the time the option is bought by the investor.

The risk associated with an option contract derives from the unknown evolution of the price of the underlying asset on the market. This risk can

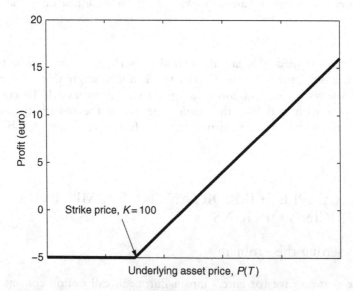

Figure 6.11 Profit (euro) gained from buying an European call option with strike price of 100 euros and premium of 5 euros

EXAMPLE 2: IMPORTANCE OF JUMPS IN PRICING OPTIONS

neither be controlled, nor avoided, and is an intrinsic feature of the contract itself (it is represented by the stochastic parts of the model). Besides this type of risk, another element of risk is due to the fact that the current option price is a quantity estimated via a mathematical model of a number of input variables whose values are affected by uncertainty. This uncertainty propagates through the model and produces uncertainty in the current option price. The questions we come to answer in this example are:

- How much is the uncertainty associated with a current option price?
- Which are the main sources of this uncertainty?
- Is it important to consider the effects of jumps in modelling the underlying dynamics?

The model chosen for describing the evolution of $P = \{P_t, t \geq 0\}$ over time is the Heston stochastic volatility model (Heston, 1993) with jumps (Bakshi *et al.*, 1997). Volatility behaves stochastically over time and jumps are included in the dynamics of the asset. In the analysis, we will study how uncertainties in the input factors propagate through to the output using different scenarios that correspond to combinations of different strike prices and times to maturity.

6.2.2 The Heston Stochastic Volatility Model with Jumps

Heston (1993) proposes a stock price model which introduces stochastic volatility in the famous Black–Scholes model (Black and Scholes, 1973). In the Heston model the price paths of a stock (and the volatility) are continuous. Later, extensions of the Heston model were formulated that allowed for jumps in the stock price paths (Bakshi *et al.*, 1997). We investigate these extensions and ask ourselves whether the introduced jumps play a significant role in explaining the variance of the prices of derivatives under this model.

In the following we will assume that the underlying asset is a stock with price process denoted by $P = \{P_t, t \geq 0\}$ and assume that the stock pays out a continuous compound dividend yield of $q \geq 0$. Moreover, we assume that in our market we have a risk-free bank account at our disposal, paying out a continuous risk-free interest rate r. In the Heston stochastic volatility model with jumps the stock price process is modelled by a stochastic differential equation given by

$$\frac{dP_t}{P_t} = (r - q - \lambda \mu_J)dt + \sigma_t dW_t + J_t dN_t, \quad P_0 \geq 0 \qquad (6.4)$$

where P_0 is price at time zero and $N = \{N_t, t \geq 0\}$ is a Poisson process with intensity parameter $\lambda > 0$, i.e. $E[N_t] = \lambda t$. J_t is the percentage jump size (conditional on a jump occurring) that is assumed to be log-normally, identically and independently distributed over time, with unconditional mean μ_J. The standard deviation of $\log(1+J_t)$ is σ_J:

$$\log(1+J_t) \sim N\left(\log(1+\mu_J) - \frac{\sigma_J^2}{2}, \sigma_J^2\right) \tag{6.5}$$

The (squared) volatility follows the classical Cox–Ingersoll–Ross (CIR) process:

$$d\sigma_t^2 = k(\eta - \sigma_t^2)dt + \theta\sigma_t d\tilde{W}_t, \; \sigma_0 \geq 0 \tag{6.6}$$

where $W = \{W_t, t \geq 0\}$ and $\tilde{W} = \{\tilde{W}_t, t \geq 0\}$ are two correlated standard Brownian motions such that $\text{Cov}[dW_t, d\tilde{W}_t] = \rho dt$. Finally, J_t and N are independent, as well as W_t and $\tilde{W}_t$. The parameter η is interpreted as the long-run squared volatility; κ is the rate of mean reversion to the level η and θ governs the volatility of volatility. The characteristic function $\phi(u,t)$ of the logarithm of the stock price process for this model is given by

$$\begin{aligned}\phi(u,t) &= E\{\exp(iu\log(P_t))\,|P_0, \sigma_0^2\} \\ &= \exp(iu(\log P_0 + (r-q)t)) \\ &\quad \times \exp(\eta\kappa\theta^{-2}((\kappa - \rho\theta ui - d)t - 2\log((1-ge^{-dt})/(1-g)))) \\ &\quad \times \exp(\sigma_0^2\theta^{-2}((\kappa - \rho\theta ui - d)(1-e^{-dt})/(1-ge^{-dt}))) \\ &\quad \times \exp(-\lambda\mu_J tui + \lambda t((1+\mu_J)^{ui}\exp(\sigma_J^2(ui/2)(ui-1)) - 1)) \end{aligned} \tag{6.7}$$

where

$$d = ((\rho\theta ui - \kappa)^2 - \theta^2(-ui - u^2))^{1/2},$$

$$g = (\kappa - \rho\theta ui - d)/(\kappa - \rho\theta ui + d).$$

Pricing of European call options under this model can be done by the Carr and Madan (1998) pricing method which is applicable when the characteristic function of the logarithm of the risk-neutral stock price process is known. Let α be a positive constant such that the αth moment of the stock price exists. Carr and Madan (1998) showed that the price $C(K, T)$ at time $t = 0$ of a European call option with strike K and time to maturity T is given by

$$C(K, T) = \frac{\exp(-\alpha\log(K))}{\pi}\int_0^{+\infty}\exp(-iv\log(K))\rho(v)dv$$

EXAMPLE 2: IMPORTANCE OF JUMPS IN PRICING OPTIONS

where

$$\rho(v) = \frac{\exp(-rT)E[\exp(i(v-(\alpha+1)i)\log(P_T))]}{\alpha^2+\alpha-v^2+i(2\alpha+1)v}$$

$$= \frac{\exp(-rT)\phi(v-(\alpha+1)i, T)}{\alpha^2+\alpha-v^2+i(2\alpha+1)v}. \tag{6.8}$$

Using fast Fourier transforms, one can compute rapidly the complete option surface with an ordinary computer.

The input factors selected for sensitivity analysis purposes are listed in Table 6.6. They can be differentiated into two groups: those whose value can be estimated with a certain degree of confidence using market data, and therefore represent a source of uncertainty that can be defined as 'controllable'; and those that cannot be checked with market data and are therefore regarded as completely 'uncontrollable'. The first group consists of the initial condition for the dynamics of the volatility σ_0, the dividend yield q, and the interest rate r. The remaining inputs, among which are the jump parameters, belong to the second group.

Table 6.6 List of input factors of the Heston model with jumps. The last two columns report the lower and upper bounds of the uniform distributions for each input

	Input	Description	Minimum	Maximum
Controllable	σ_0	Initial condition for the dynamics of the volatility	0.04	0.09
	q	Dividend yield	0.00	0.05
	r	Interest rate	0.00	0.05
Uncontrollable	κ	Rate of mean revision	0.00	1.00
	η	Long-run squared volatility	0.04	0.09
	θ	Parameter governing the volatility of volatility	0.20	0.50
	ρ	Correlation between Brownian motions in Equations ((6.4)) and ((6.6))	−1.00	0.00
	λ	Jumps parameter (intensity parameter of the Poisson process that models the stock price)	0.00	2.00
	μ_j	Jump parameter in Equation ((6.5))	−0.10	0.10
	σ_j	Jump parameter in Equation ((6.5))	0.00	0.20

6.2.3 Selecting a Suitable Sensitivity Analysis Method

In total there are 10 input factors in our model that could affect the pricing of European call options. The model takes a few seconds to run and the specific answers we try to get by the application of sensitivity analysis are:

- What is the relative importance of the 10 input factors affecting the uncertainty in the current option pricing in difference scenarios, i.e. combinations of the option strike price K and of the time to maturity T?
- Does the impact of the uncertainties change from scenario to scenario?
- Is it the controllable or the uncontrollable input factors that drive most of the output variance?

Given the low number of input factors, the low computational cost of the model and the fact that we want to assess both first-order and total-effect sensitivity indices $\{S_i, S_{T_i}\}$, the most appropriate sensitivity analysis method to use is a variance-based method (see Chapter 4), as these methods do not rely on assumptions about the model being additive, or monotonic in the input–output relationship.

6.2.4 The Sensitivity Analysis Experiment and Results

The first-order S_i and total-effect S_{T_i} sensitivity indices are computed for each input factor in different scenarios. These scenarios represent different combinations of the option strike price K and the time to maturity T. The initial condition for the stock is fixed at $P_0 = 100$, while seven values of the strike price are considered ($K = 70, 80, \ldots, 130$) to represent situations in which the option is in the money ($K \leq P_0$), at the money ($K = P_0$) or out of the money ($K \geq P_0$). Six different time horizons are examined from $T = 0.5$ years up to $T = 3$ years. Therefore, we study 42 scenarios in total. The total number of model executions to estimate the entire set of the variance-based indices $\{S_i, S_{T_i}, i = 1, 2, \ldots 10\}$ is $N = 24.576 (= 2048(10+2))$.

Figure 6.12 shows the importance of the input factors in each of the 42 scenarios according to the S_{T_i} scaled in $[0, 1]$ for presentational purposes. Results reveal that, across all scenarios, the most influential parameters are: the dividend yield q, the interest rate r, and the jump parameters λ and σ_J. In particular, q and r are very important for low strike prices at all times to maturity, while λ and σ_J become more relevant at higher strike prices. As expected, σ_0 is important only for low times to maturity, especially when the option is not in the money. Note that among these four most important factors, q and r belong to the group of 'controllable' factors. More interesting are the roles of the jump parameters λ and σ_J, which are the most relevant among the 'uncontrollable' factors. In general, if we limit our attention to the 'uncontrollable' factors, the analysis proves that the

EXAMPLE 2: IMPORTANCE OF JUMPS IN PRICING OPTIONS 259

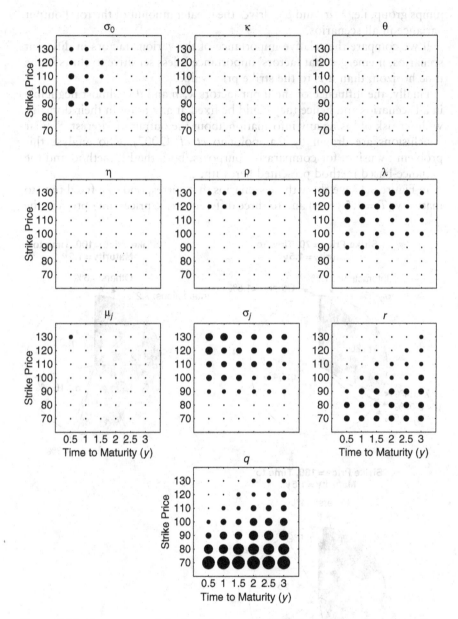

Figure 6.12 Importance of the factors in each of the 42 (6×7) scenarios according to S_{T_i} scaled in [0, 1]. Each graph refers to a selected input factor. Within a graph, each dot illustrates the input importance in a specific scenario. Ten different dots' sizes are considered, corresponding to 10 different classes of importance, ranging from values of S_{T_i} in [0, 0.1] to values of S_{T_i} in [0.9, 1]

jumps group, i.e. λ, σ_J and μ_J, drives the greater amount of the total output variance in all scenarios.

If we compare the relative importance of the various factors in different scenarios, it emerges that factors' importance is less sensitive to shifts in the time horizon, than it is to the strike price value.

Finally, the influence of the input factors κ, η and θ is almost negligible in all scenarios and hence they could be fixed at any point in their domains without risk of losing any information about the output of interest. Similar conclusions are drawn by Campolongo *et al.* (2007), who analyse this problem by using, for comparative purposes, both the EE method and the variance-based method presented herewith.

In Figure 6.13 we show three scenarios that correspond to a fixed time to maturity ($T = 1.5$ years) and to three different strike prices: an option in the

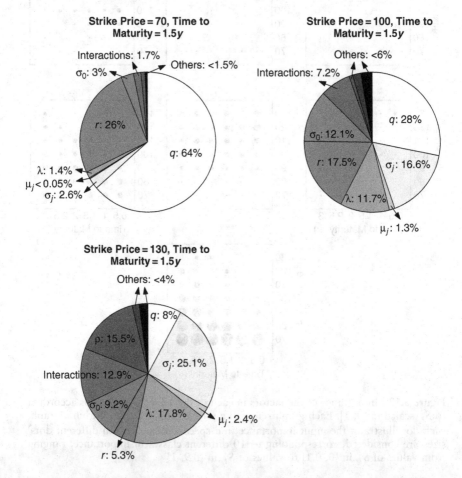

Figure 6.13 Decomposition of the total variance of the option price in three scenarios. Interactions of all orders are grouped in a single term

EXAMPLE 2: IMPORTANCE OF JUMPS IN PRICING OPTIONS

money ($K = 70$), at the money ($K = 100$) and out of the money ($K = 130$). The variance of the option price is apportioned to the contributions due to the first-order effects of each input (i.e. the Sobol' S_i) and to the interactions of all orders. Since we are not interested in quantifying interactions of different orders due to certain input factors, all interactions are summed up in a single term. The figure reveals that the importance of jumps (factors λ and σ_J) increases considerably with the strike price. The sum of their main effects goes from approximately 4%, when the option is in the money, to nearly 30% when the option is at the money, up to 45.3% when the option reaches 130 (out of the money). When the option is in the money, more than 90% of the total variance is due to the controllable factors q and r, thus leaving little opportunity to reduce the uncertainty in the option price by increasing the modelling effort. As the strike price increases, the importance of jumps increases considerably, which stresses that their role in modelling the option price cannot be overlooked: jumps need to be included in the model and their representation should be as accurate as possible. We could summarize in two main points the impact of these findings on the investor's actions:

- Among the important factors, q and r are controllable. Therefore, the investor might reduce the uncertainty in $C(K, T)$ by prioritizing his efforts to get good estimates of q and r from market data.
- Among the uncontrollable factors, it is the jumps that drive most of the variance. However, this type of uncertainty remains in the model and there is no possibility for the investor to reduce it. But, from a modelling point of view, the analysis presented above has shown that the inclusion of jumps is of paramount importance in the estimation of the underlying asset price.

6.2.5 Conclusions

In this example we have analysed the uncertainty in the price of a European option estimated by the Heston model in its version with jumps. The uncertainty in the option price has been apportioned to its different sources via sensitivity analysis. In particular we have assessed the relative importance of jumps with respect to other model input factors.

The sensitivity analysis using variance-based sensitivity indices S_i, S_{T_i} has shown that, among the 'uncontrollable' factors (i.e. those that cannot be estimated from market data), jumps play a major role in determining the option price, thereby emphasizing the need to include them in the model formulation. At low strike prices, most of the uncertainty in the option price is due to 'controllable' factors such as the dividend yield q and the interest rate r. As the option strike price increases, the importance of jumps increases considerably: for instance, for options with strike prices of 120 or 130 (the starting price being 100), the importance of jumps is superior to that of q and r for all times to maturity. This underlines that an accurate assessment of the jump process becomes more urgent for out-of-the-money options. It also emerges that three of the inputs, namely the rate of mean reversion κ, the long-run squared volatility η and the parameter governing the volatility of volatility θ, do not affect the output in any of the studied scenarios. This result is useful, as it may allow a simplification of the model. Finally, as expected, at low time to maturity the initial condition for volatility needs to be accurately determined, while it becomes less important as the time to maturity increases.

6.3 EXAMPLE 3: A CHEMICAL REACTOR

6.3.1 Setting the Problem

Imagine that a chemist, in a routine procedure, attempts to create a model of reaction systems to understand reaction mechanisms, kinetic properties, process yields under various operating conditions, or the impact of chemicals on humans and the environment. These models are attempts to mimic the system by hypothesizing, extracting and encoding system features (e.g. a potentially relevant reaction pathway versus another plausible one). A model will hopefully help to corroborate or falsify a given description of reality, e.g. by validating a reaction scheme for a photochemical process in the atmosphere, and may also help to influence it, e.g. by facilitating the identification of optimal operating conditions for an industrial process or by suggesting mitigating strategies for an undesired environmental impact.

These models are customarily built in the presence of uncertainties of various levels, in the pathway, in the order of the kinetics associated with the pathway, in the numerical value of the kinetic and thermodynamic constants for that pathway and so on. The job of the uncertainty analysis is to propagate all these uncertainties, via the model, onto the model output of interest, e.g. the yield of a process. The work of sensitivity analysis is to determine the strength of the relation between a given uncertain input and the output.

The issues a chemist may want to deal with during this modelling process could include:

- understanding the reaction path, mechanism, or rate-determining steps in a detailed kinetic model with a large number of elementary reactions (Turanyi, 1990; Saltelli and Hjorth, 1995);

EXAMPLE 3: A CHEMICAL REACTOR

- extracting important elementary reactions from a complex kinetic model to obtain a reduced model (e.g. a minimal reaction scheme) with equivalent predictive power;
- selecting important reactions for further analysis (Pandis and Seinfeld, 1989; Vuilleumier et al., 1997);
- estimating the output of a computer program in the neighbourhood of a given set of boundary conditions without rerunning the program (Grievank, 2000; Kioutsioukis et al., 2005);
- reconciling model parameters with observations (Rabitz, 1989; Le Dimet et al., 2002; Mallet and Sportisse, 2004).

6.3.2 Thermal Runaway Analysis of a Batch Reactor

The thermal runaway analysis of a batch reactor, where an exothermic reaction $A \to B$ takes place, is performed considering the following mass and heat balance equations:

$$\frac{d[A]}{dt} = -k(T) \cdot [A]^n \qquad (6.9)$$

$$\rho c_v \frac{dT}{dt} = (-\Delta H) \cdot k(T) \cdot [A]^n - s_v \cdot u \cdot (T - T_a) \qquad (6.10)$$

where n is the order of the reaction, ρ the density of the fluid mixture [kg/m^3], c_v the mean specific heat capacity of the reaction mixture [J/(Kmol)], ΔH the reaction enthalpy [J/mol], s_v the surface area per unit volume [m^2/m^3], u the overall heat transfer coefficient [J/(m^{2}s K)] and T_a the ambient temperature [K]. The initial conditions are $[A] = [A]_0$, $T = T_0$, $t = 0$. This is customarily rewritten in dimensionless form:

$$\frac{dx}{d\tau} = \exp\left(\frac{\theta}{1+\theta/\gamma}\right) \cdot (1-x)^n = F_1(x, \theta)$$

$$\frac{d\theta}{d\tau} = B \cdot \exp\left(\frac{\theta}{1+\theta/\gamma}\right) \cdot (1-x)^n - \frac{B}{\psi} \cdot (\theta - \theta_a) = F_2(x, \theta)$$

with initial conditions $x = 0$, $\theta = 0$ at $\tau = 0$, and the dimensionless variables:

$$x = \frac{[A]_0 - [A]}{[A]_0} \qquad \theta = \frac{T_0 - T}{T_0} \cdot \gamma \qquad \tau = t \cdot k(T_0) \cdot ([A]_0)^{n-1}$$

and dimensionless parameters:

$$B = \frac{(-\Delta H) \cdot [A]_0}{\rho \cdot c_v \cdot T_0} \gamma \quad \text{(heat of reaction)}$$

$$\gamma = \frac{E}{R_g T_0} \quad \text{(activation energy)}$$

$$\psi = \frac{(-\Delta H) \cdot k(T_0) \cdot ([A]_0)^n}{s_v \cdot u \cdot T_0} \cdot \gamma \quad \text{(Semenov number expressed as}$$
(heat release potential)/(heat removal potential)

This system was widely analysed in the last century to characterize thermal runaway under varying operating conditions (Varma et al., 1999). At a given constant rate and ambient temperature, the system is completely determined by the parameters B and ψ, and critical conditions are usually illustrated in the B–ψ plane.

A reactor under explosive conditions is sensitive to small variations in, for example, the initial temperature, while under nonexplosive conditions, the system remains insensitive to such variations. The boundaries between runaway (explosive) and nonrunaway (nonexplosive) conditions can thus be identified based on the system's sensitivity to small changes in the operating parameters. The system can also be characterized by the derivative of the maximum temperature reached in the reactor versus the initial temperature, i.e. $S(\theta^*, \theta_0) = d\theta^*/d\theta_0$ (Morbidelli and Varma, 1988).

The runaway boundary is defined as the critical value of each parameter for which the sensitivity to the initial condition is at a maximum, e.g. for the Semenov number ψ we have the results in Figure 6.14.

For ψ values smaller than ψ_c the system is in nonrunaway conditions, i.e. the maximum temperature reached in the reactor is not very high and this maximum is insensitive to small variations in the inlet temperature. Increasing ψ, both the maximum temperature and its sensitivity to T_0 increase smoothly until, in proximity to ψ_c, there is a sharp rise for both of them that rapidly brings the reactor to a strong temperature increase. For ψ values higher than ψ_c, the sensitivity decreases again, leaving unchanged the extreme temperature rise reached at ψ_c. By fixing the reaction kinetics (n, γ) and the ambient temperature (θ_a), the curve in the B–ψ plane can be obtained (Figure 6.15).

Consider the case of a system with nominal parameter design $B = 20$, $\gamma = 20$, $n = 1$ and $\psi = 0.5$. Under these conditions the system should be in the nonrunaway region. The system, however, is characterized by uncertainties. Let us therefore assume the following uncertainty distributions for the model parameters:

EXAMPLE 3: A CHEMICAL REACTOR

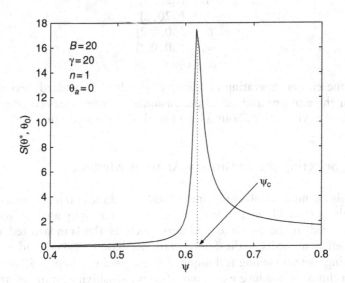

Figure 6.14 $S(\theta^*, \theta_i) = d\theta^*/d\theta_i$ versus Semenov number ψ for the model expressed by Equations (6.9) and (6.10)

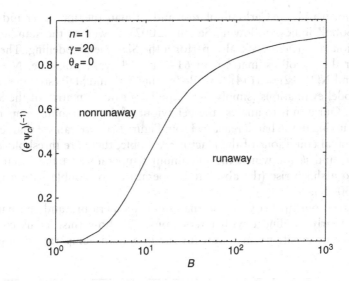

Figure 6.15 Runaway versus nonrunaway in the plane B, $1/e\psi$ for fixed n, γ and θ_a (e is the number of Neper)

$$\begin{aligned} B &\sim N(20, 4) \\ \gamma &\sim N(20, 2) \\ \theta_a &\sim N(0, 0.2) \\ \theta_0 &\sim N(0, 0.2) \\ \psi &\sim U(0.4, 0.6) \end{aligned} \qquad (6.11)$$

Under the chosen operating conditions ($\gamma = 20$), a standard deviation of 0.02 for the ambient and initial dimensionless temperatures correspond to a standard deviation of about 3 K in the absolute temperature scale.

6.3.3 Selecting the Sensitivity Analysis Method

Since this example involves few input factors and the computational cost of the analysis is small, we employ several of the sensitivity analysis methods we discussed in the previous chapters, such as the standardized regression coefficients (SRC), the Sobol' variance-based method and the SDP smoothing metamodelling technique. As discussed in Chapter 5, metamodelling techniques produce extremely efficient sensitivity estimates and will be considered as the 'reference' values in the following example.

6.3.4 The Sensitivity Analysis Experiment and Results

We perform a Monte Carlo simulation and estimate the first-order and total-effect Sobol' indices following Saltelli (2002) as well as the standardized regression coefficients. We also perform the SDP metamodelling. The total cost for the Sobol' estimates is of 6144 model evaluations (i.e. $N = 512$, $k = 5$ and $N(2k+2) = 6144$),[4] while for the SRC and SDP estimates it is of 512 model evaluations (simply using the base sample matrix of the Sobol' design). Our aim is to analyse the behaviour of the maximum temperature rise occurring in the batch reactor. From Figure 6.16 we can see that, even if the nominal conditions of the reactor are stable, there are threshold values of B, θ_a and ψ for which the maximum temperature in the reactor can undergo a sharp rise (the absolute temperature can double with a rise of about 300 K).

The relationship between the maximum temperature and the parameters is clearly nonlinear with interactions. To show this, let us consider the different sensitivity measures for the maximum temperature shown in

---

[4] This is not the recipe given in Chapter 4, which would require $N(k+2) = 3584$ model evaluations, but corresponds to the original design of Sobol'. This larger sample design has been used because, as discussed in Saltelli (2002), it allows us also to obtain estimates of second-order sensitivity indices, which are discussed later in this example.

EXAMPLE 3: A CHEMICAL REACTOR

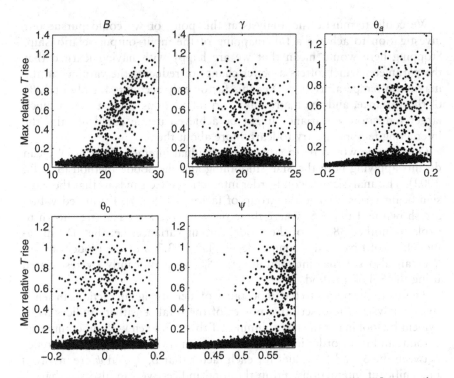

Figure 6.16 Relative temperature rise at the maximum $[(T_{max} - T_0)/T_0]$ versus the uncertain model parameters. The extreme values of about 1.2 for $[(T_{max} - T_0)/T_0]$ correspond to a temperature rise of about 300 K

Table 6.7 Sensitivity measures for model ((6.9))–((6.11))

	β_i^2	S_i (Sobol')	S_i (SDP)	S_{T_i}
ψ	0.17	0.1781	0.205	0.6738
θ_a	0.14	0.1641	0.148	0.556
B	0.087	0.08	0.092	0.4692
γ	0.0012	0.0019	0.001	0.0322
θ_0	0.0006	0.0015	0.0004	0.0128
Sum	0.40	0.43	0.45	

Table 6.7. Consider the β_i^2, S_i columns first. The sensitivity based on the β_i^2's can only capture 40% of the variation of the maximum temperature. Taking the SDP estimate of the main effects into account, we get 45% (43% using the Sobol' method). This leaves 55% to interactions between the model parameters. Results in Table 6.7 also show that Sobol' estimates are reasonably close to the 'reference' SDP estimates.

We could terminate the analysis at this point or we could pursue our investigation to achieve a full mapping of the input–output relationship. Stopping here would mean that we are happy with having learned that the parameter which offers a better chance of reducing the variance in the maximum temperature is ψ. Yet this factor only accounts for about 20% of the variance, and the many unknown interactions suggest that a much larger reduction in variance might be achieved if one could identify the interacting factors and try to learn more about them.

One avenue would be to compute individual interaction terms. We can do this applying both the SDP smoothing and the Sobol' method (Saltelli, 2002). The analysis of second-order interaction effects tells us that the most significant terms involve the group of factors (B, θ_a, ψ). Estimated values are shown in Table 6.8, where these three interaction terms are shown to explain another 38.7% of the model output variance, i.e. the SDP meta-modelling of first- and second-order explains 83.7% of the model output. We can also see that interaction term $S_{B\theta_a}$ is significantly overestimated using the Sobol' method.

This example points to the importance of identifying interactions in sensitivity analysis. The descriptive power of the total sensitivity indices S_{T_i} is evident by looking at the last column of Table 6.7. Even without computing second- and third-order interaction effects, it is evident from the difference between the S_i and S_{T_i} values for each factor that B, θ_a and ψ are involved in significant interactions. From the total indices we can also see that all the interaction terms of factor γ with (B, θ_a, ψ) cover about 3% of the total variance, while θ_0 has a negligible total effect, implying its irrelevance.

In Figure 6.17 we show the SDP estimates of first-order ANOVA-HDMR terms of the maximum relative temperature rise $(T^{\max} - T_0)/T_0$ for the three most important parameters. These show that the first-order relationships are monotonic, which explains why the β_i^2 give an acceptable estimate of the sensitivities in Table 6.7, with an average error of about 0.05.

In Figure 6.18 we also show the SDP estimates of the second-order ANOVA-HDMR terms for the group of factors (B, θ_a, ψ). It can be clearly seen that extreme values of the temperature rise are associated with high–high combinations of each couple of factors. In particular, the highest peaks are obtained combining high values for B and ψ.

Table 6.8 Second-order sensitivity indices for model ((6.9))–((6.11))

	S_{ij} (Sobol')	S_{ij} (SDP)
$S_{B\psi}$	0.17	0.169
$S_{\theta_a \psi}$	0.17	0.144
$S_{B\theta_a}$	0.166	0.074

EXAMPLE 3: A CHEMICAL REACTOR

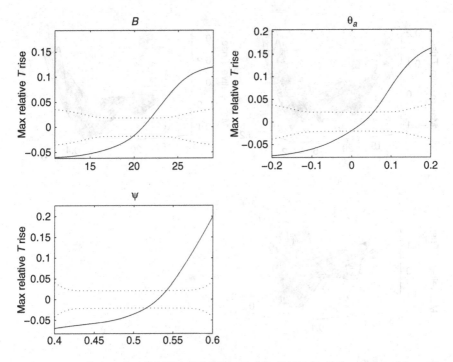

Figure 6.17 First-order ANOVA-HDMR terms (solid lines) of the subset of factors (B, θ_a, ψ) for the maximum relative temperature rise, estimated using the SDP procedure described in Chapter 5 (note: the mean value of the maximum temperature rise $(T_{max} - T_0)/T_0$ is 0.1). Dotted lines show the 95% standard error band

6.3.5 Conclusions

In the analysis of thermal runaway of a batch reactor, the sensitivity analysis results can be summarized as follows:

- The parameter which offers better chances of reducing the variance in the maximum temperature is ψ (about 20% of the variance). The sensitivity based on the $\beta'_i s$ can capture only 40% of the variation of the maximum temperature. Considering the SDP estimate of the main effects, we can arrive at 45%. This implies that 55% is caused by interactions between the five model parameters.
- The most significant second-order interaction terms are: $S_{B\psi} = 0.169$, $S_{\theta_a \psi} = 0.14$, $S_{B\theta_a} = 0.074$. The SDP estimates also suggest that high–high combinations of such input factors are most responsible for extreme temperature rises. Joining these results with first-order terms, covers about 84% of the total output variance. The remaining part can be

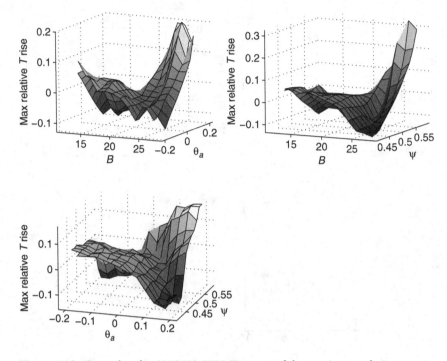

Figure 6.18 Second-order ANOVA-HDMR terms of the maximum relative temperature rise, estimated with the SDP procedure described in Chapter 5

attributed to higher interactions involving the most relevant group of factors (B, θ_a, ψ) as well as γ which has a nonnegligible total effect.
- The first-order ANOVA-HDMR terms of the 'maximum relative temperature rise' for the three most important parameters show that the first-order relationships are monotonic, which explains why β_i^2 give an acceptable estimate of the sensitivities.

6.4 EXAMPLE 4: A MIXED UNCERTAINTY–SENSITIVITY PLOT

6.4.1 In Brief

As a last example, we treat a mixed uncertainty–sensitivity plot that might be meaningful to particular audiences and which could be used to convey information on the impact of changes in an influential factor on an output of interest. This resembles the one-at-a-time (OAT) type of sensitivity analysis discussed in Chapter 3.

EXAMPLE 4: A MIXED UNCERTAINTY–SENSITIVITY PLOT

The example is taken from a study on clearing and settlement (Schulze and Baur, 2006, p. 18), published on the European Commission Directorate General Internal Market website. Without entering into the details of the model, we focus on the impact of a single input factor (i.e. the trading costs) on an output of interest (i.e. GDP in the European Union). To demonstrate that trading costs are highly influential on GDP, the authors calculated the changes in GDP due to changes in trading costs, considering different mean values for the trading costs in the range 0–30% (in steps of 2.5%). A coefficient of variation of 30% is assumed for each mean value. Their plot, presented in Figure 6.19, showed that GDP increases strongly with higher reductions in trading costs, but so do the confidence bounds. For example, the average change in the level of GDP in the case of a 7% reduction in trading costs is 23 billion (with a 95% confidence interval: [8 billion, 47 billion]). Higher reductions in trading costs, e.g. of about 18%, could lead to a change in the level of GDP of 63 billion (95% confidence interval: [20 billion, 124 billion]).

This type of graph can serve a dual purpose. On one hand, it can be used to present the results of an uncertainty analysis, in which confidence bounds are estimated for an output of interest, such as GDP. On the other hand, it can show the results of a sensitivity analysis, revealing how a crucial uncertain factor (the percentage reduction in trading cost) can clearly

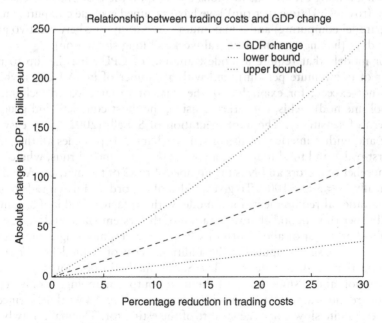

Figure 6.19 A mixed uncertainty–sensitivity analysis plot

influence the output of interest (GDP), taking account of the uncertainty in the other factors. Though very simple, this analysis helped the stakeholders to hold a well-structured negotiation on the estimated benefit of a possible reduction in trading cost.

6.5 WHEN TO USE WHAT?

The choice of the proper sensitivity analysis technique depends on such considerations as:

- the computational cost of running the model;
- the number of input factors;
- features of the model (e.g. linearity);
- the consideration of interactions among the input parameters in the model;
- the setting for the analysis and its audience.

Table 6.9 summarizes the explanations using these characteristics as discriminating criteria. These indications are not to be taken as a prescription. More than one method might be applicable to the same context. In choosing among the various methods based upon the cost of a single simulation (row: 'CPU time (per run)') we have assumed a single computing unit (no parallel computing) and a maximum computing time between two and three days (having in mind a typical weekend-long simulation).

For models that require a modest amount of CPU time (i.e. up to the order of one minute per run), and with a number of input factors which does not exceed, for example, 20, the class of variance-based techniques (Sobol' method) yields, convergence aside, the most complete and general pattern of sensitivity. The implementation of Saltelli (2002) (very easy to code and with numerical recipes given in Chapter 4) provides all the pairs of first-order and total indices at a cost of $(k+2)N$ model runs, where k is the number of factors and N is the number of rows of the matrices $\mathbf{A}$ and $\mathbf{B}$. Typically $N \approx 500 \div 1000$. To give an idea of the order of magnitude of the computational requirement: for a model with 10 factors and half a minute of CPU per run, a good characterization of the system via S_i and S_{T_i} can be obtained at a cost of about 50 h of CPU. Moreover, any single interaction term can be easily computed at the additional cost of N model evaluations per sensitivity index, if desired. The Sobol' method does not rely on any assumption about smoothness of the input–output mapping; it only relies on square-integrability of Y. This is both a strength and a weakness, since it implies a quite slow convergence rate of the estimator. The main drawback of this method is therefore in the computational cost, which depends on

Table 6.9 When to use what. Note that these indications are not to be taken as a prescription. More than one method might be applicable to the same context. In choosing among the various methods based upon the cost of a single simulation (row: CPU per run) we have assumed a single computing unit (no parallel computing) and a maximum computing time between two and three days (having in mind a typical weekend-long simulation)

Characteristic	Chapter 1	Chapter 1	Chapter 2	Chapter 3	Chapter 4	Chapter 5	Chapter 5
	β_i^2	Scatterplots	Fractional factorial	Elementary effects	Variance based	Metamodelling	Monte Carlo filtering
Coping with nonlinearity	No[a]	Yes	Yes	Yes	Yes	Yes	Yes
Coping with interactions	No	Yes	Yes	Yes	Yes	Yes	Yes
Samples taken from	Distributions	Distributions	Levels	Levels	Distributions	Distributions	Distributions
Number of input factors	<100	<10	>100[b]	20–100[b]	<20[b]	20–100	<20
CPU time per run	1 min–1h	<1h	<10h	<1h	<1 min	<1h	<1h
Cost of analysis (number of runs)	500–1000	1000	$k \div 2k$ (FF)	$r(k+1)$	$N(k+2)$ (Saltelli, 2002) N (RBD)	100–1000	500–2000
Setting for SA	FP	FM	$N<k$ (supersat. FF) FF	FF	FP, FF, VC	FP, FM	FM

k: number of factors.
N: typically $N \approx 500 \div 1000$.
r is generally set to $r \approx (4 \div 10)$ and corresponds to the number of the trajectories.
FP: Factor prioritization; FF: Factor fixing; VC: Variance cutting; FM: Factor mapping.
Notes:
[a] Recommended when $R^2 \geq 0.7$. In its ranked transformed version it can be quite effective for *monotonic* models, irrespective of their degree of nonlinearity.
[b] Treating *groups* of factors would allow to treat problems of higher dimensionality.

the dimension k of the problem and in the relatively large N required for a reasonable accuracy of the sensitivity estimates.

When using Saltelli's recipe, we make use of quasi-random numbers to generate the sample matrices **A** and **B** (see Chapter 4) for the analysis. These are sequences of multidimensional points characterized by the property of 'optimal' space filling (see Chapter 2). When the input factors are correlated an ad hoc computational scheme must be adopted (see the brief discussion in Chapter 1).

The method based on random balance design (RBD) provides only first-order indices. However, the method has the advantage of running a full analysis at a cost of N model runs, i.e. independent on the number of input factors k. At a given computational cost, RBD yields first-order estimates at higher accuracy than Saltelli (2002) (see comparison tests in Tarantola et al., 2006). In addition, RBD is very easy to code (see the few-line Matlab script in Chapter 4).

A cheaper alternative to the variance-based methods are the standardized regression coefficients, β_i. With a single batch of N sampled points (e.g. $N \approx 500 \div 1000$ points or fewer depending on the cost of the model) the β_i and their rank transformed version can be estimated for all the input factors. The β_i are only effective for linear or quasi-linear models, i.e. for $R^2 \geqslant 0.7$. Regression methods, also implemented in SIMLAB, are always useful to look at when investigating the degree of linearity of the model.

As discussed in Chapter 5, metamodelling techniques, like the SDP approach, can be applied successfully to produce both truncated ANOVA-HDMR decompositions up to third order as well as the relative sensitivity indices at the same cost of β_i. This strongly reduces the cost of the sensitivity analysis with respect to Saltelli (2002), with extremely reliable estimates of the full set of main effects at very small sample sizes ($N = 250 \div 500$). The strongly improved convergence rate of metamodelling techniques with respect to the classic Sobol' method is due to their reliance of smoothness assumptions about Y. However, accurate estimates of the total effect are problematic in the metamodelling approach, unless the model has a small number of interaction terms of very small order, i.e when the truncated ANOVA-HDMR covers 100% of the model output Y.

When the CPU time increases (up to, for example, 10 minutes per run), or the number of factors increases (e.g. up to 100), the EE method offers the best result. The number of sampled points required is $r(k+1)$ where r is generally set to $r \approx (4 \div 10)$ and k the number of input factors. To give an example, with 80 factors and 5 minutes CPU time per run, all the model outputs can be ready in 27 hours if $r = 4$ is taken. The EE method produces good proxies to variance-based total effects, thus allowing us to efficiently complement a metamodelling exercise.

When the number of input factors and/or the CPU time is even so large as to preclude the use of a screening method, then supersaturated fractional

WHEN TO USE WHAT?

factorial designs, where factors are iteratively perturbed in batches, can be used (see the discussion in Chapter 2). However, these methods preclude an effective exploration of the space of the inputs, as they mostly operate at very few factor levels and require strong assumptions of the model's behaviour.

Sensitivity analysis is also driven by the setting. When the purpose of the analysis is to prioritize factors, the first-order sensitivity indices S_i (or the standardized regression coefficients) have a good argument. If the objective is to fix noninfluential factors, then the total sensitivity indices S_{T_i}, or equivalently the EE method, come into use. If a particular region in the space of the output (e.g. above or below a given threshold) is of interest, then Monte Carlo filtering and associated methods can be tried alternatively or complementarily to the measures just mentioned. If the purpose of the analysis is a detailed mapping of the input–output relationship, then various types of metamodelling techniques can be successfully applied.

The reader of the present textbook may expect new developments in the computation of the important measure and screening texts described here. Ongoing research both on sampling strategies (thereby using true model points to map the space of the input) as well as metamodel/emulators (producing estimates of model output at untried points) is being intensively carried out by practitioners.

Afterword

The authors of the book hope that readers will find the tools of global sensitivity analysis presented here useful. As discussed in the introduction, these methods have been conceived to help avoid type II errors resulting from a poor (or wholly omitted) exploration of the space of the input factors or from neglecting the interaction between different system factors and/or compartments.

We have also stressed the need to consider more than just the uncertain factors or the measurement errors in the input data included in the model. Although the tools discussed in the book are capable of handling what we could call 'technical' uncertainty, they can also be used to tackle additional layers of epistemic uncertainty.

A model, being a human representation of a given problem, necessarily reflects the perceptions, values and interests of those concerned with structuring the issue being modelled. Whenever relevant, a sensitivity analysis should incorporate these social and cultural dimensions. Post-Normal Science (PNS), as discussed in the introduction, offers some guidance as to when 'extended peer community' methods are advisable.

Clearly, modelling is subject to an unprecedented level of critique at present, not only from postmodern philosophers but also from intellectuals and scientists of diverse political tendencies. In his novel *State of Fear*, Michael Crichton writes:

> If you read some author who says 'We find that anthropogenic greenhouse gases and sulphates have had a detectable influence on sea-level pressure' it sounds like they went into the world and measured something. Actually, they just run a simulation. They talk as if simulations were real-world data. They're not. That's a problem that has to be fixed. I favor a stamp: **WARNING: COMPUTER SIMULATION – MAY BE ERRONEOUS AND UNVERIFIABLE.** Like on cigarettes [···]. (Crichton, 2004, p. 556)

It is indeed a sign of the times that statements about the veracity of computational models can now be found on books belonging more on our bedside tables than on our desks.

A common fallacy in the use of computation models is – in the opinion of the authors – the idea that increasing the level of detail of a model must in general improve its pedigree.

An example discussed in Pilkey and Pilkey-Jarvis (2007) concerns the Yucca Mountain repository for radioactive waste disposal, where a very large model called TSPA (Total System Performance Assessment) is employed to guarantee the safe containment of waste. Composed of 286 submodels, TSPA (as with any other model) is based on assumptions – a crucial one being the low permeability of the geological foundation and hence the extensive period of time needed for the water to percolate from the desert surface to the level of the underground disposal. However, the confidence of stakeholders in TSPA was not strengthened by the recent production of evidence to suggest that an upward revision of the value of this parameter by four orders of magnitude might be necessary.

Many people will take exception to the works just cited. Yet it cannot be denied that a paradigm shift has taken place. As discussed in Chapter 1, stakeholders and media alike will increasingly tend to suspect, or even *expect*, the instrumental use of computational models and the amplification or suppression of uncertainty for the sake of expedience (Michaels, 2005).

A global sensitivity analysis might serve variously the friends *and* the foes of a given model as applied to a particular issue, that is, both the modellers and the critics of the model. Yet there is no magic formula – a sensitivity analysis may reveal to which assumptions a model is more sensitive but say nothing about the defensibility of these assumptions. Modellers could consider that 'the uncertainties which are more carefully scrutinized are usually those which are the least relevant', (Van der Sluijs et al., 2005). Nassim Nicholas Taleb, in an interesting book also critical of the modelling community, calls this 'The delusion of uncertainty' (Taleb, 2007). In other words, sensitivity analysis may not guard against type III errors, that is, a wrong framing of the analysis.

Have models fallen from grace, and is modelling – when applied to major environmental issues of the PNS type – just 'useless arithmetic', as claimed by Pilkey and Pilkey-Jarvis (2007)?

The crux of the matter is that, 15 years after Konikof and Bredehoeft proclaimed that 'groundwater models cannot be validated' (Konikov and Bredehoeft, 1992) and Naomi Oreskes wrote her sober reminder of the limits to validation of computational models in environmental systems (Oreskes et al., 1994), the modelling community still lacks a set of accepted quality criteria for assessing model's adequacy, notwithstanding valuable attempts at establishing one (Van der Sluijs et al., 2005).

Bibliography

Bakshi, G., C. Cao and Z. Chen (1997). Empirical performance of alternative option pricing models. *The Journal of Finance* **LII**(5), 2003–2049.

Bard, Y. (1974). *Nonlinear Parameter Estimation*. New York: Academic Press.

Beck, B. (1987). Water quality modelling: a review of the analysis of uncertainty. *Water Resources Research* **23**, 1393–1442.

Beck, M., J. Ravetz, L. Mulkey and T. Barnwell (1997). *Stochastic Hydrology and Hydraulics* **11**, 229–254.

Bell, D., H. Raiffa and A. Tversky (eds) (1988). *Decision Making: Descriptive, Normative and Prescriptive Interactions*. Cambridge University Press.

Beven, K. (1993). Prophecy, reality and uncertainty in distributed hydrological modelling. *Advances in Water Resources* **16**, 41–51.

Beven, K. (2001). *Rainfall-Runoff Modelling: The Primer*. Chichester: John Wiley & Sons, Ltd. See also Saltelli *et al.* (2000), pp. 151–192.

Black, F. and M. Scholes (1973). The pricing of options and corporate liabilities. *Journal of Political Economy* **81**, 637–654.

Borgonovo, E. (2006). Measuring uncertainty importance: investigation and comparison of alternative approaches. *Risk Analysis* **26**, 1349–1361.

Borgonovo, E. and G. Apostolakis (2001). A new importance measure for risk-informed decision making. *Reliability Engineering and System Safety* **72**, 193–212.

Box, G., W. Hunter and J. Hunter (1978). *In Statistics for Experimenters*. New York: John Wiley & Sons.

Bryson, A. and Y. Ho (1969). *Applied Optimal Control*. Blaisdell, Waltham, Mass.

Cacuci, D. (2003). *Sensitivity and Uncertainty Analysis*, Volume 1: *Theory*. Chapman and Hall.

Campolongo, F., J. Cariboni and A. Saltelli (2007). An effective screening design for sensitivity analysis of large models. *Environmental Modelling and Software* **22**, 1509–1518.

Campolongo, F., J. Kleijnen and T. Andres (2000). Screening methods. In A. Saltelli, K. Chan and M. Scott (eds), *Sensitivity Analysis*, pp. 65–80. New York: John Wiley & Sons, Ltd.

Campolongo, F. and A. Saltelli (1997). Sensitivity analysis of an environmental model: a worked application of different analysis methods. *Reliability Engineering and System Safety* **52**, 49–69.

Campolongo, F., A. Saltelli, N. R. Jensen, J. Wilson and J. Hjorth (1999a). The role of multiphase chemistry in the oxidation of dimethylsulphide (dms): a latitude dependent analysis. *Journal of Atmospheric Chemistry* **32**, 327–356.

Campolongo, F., S. Tarantola and A. Saltelli (1999b). Tackling quantitatively large dimensionality problems. *Computer Physics Communications* **117**, 75–85.

Carr, P. and D. Madan (1998). Option valuation using the fast Fourier transform. *Journal of Computational Finance* **2**, 61–73.

Chatfield, C. (1993). Model uncertainty, data mining and statistical inference. *Journal of the Royal Statistical Society A* **158**(3), 419–466.

Colbourn, C. and J. Dinitz (eds) (1996). *CRC Handbook of Combinatorial Designs*. CRC Press.

Craye, M., S. Funtowicz, P. Kloprogge, J. Ravetz and J. Risbey (2005). Combining quantitative and qualitative measures of uncertainty in model based environmental assessment: the nusap system. *Risk Analysis* **25**(2), 481–492.

Crichton, M. (2004). *State of Fear*. New York: HarperCollins.

Cukier, R. I., C. M. Fortuin, K. E. Schuler, A. G. Petschek and J. H. Schaibly (1973). Study of the sensitivity of coupled reaction systems to uncertainties in rate coefficients. i theory. *The Journal of Chemical Physics* **59**, 3873–3878.

Cukier, R., H. Levine and K. Shuler (1978). Nonlinear sensitivity analysis of multiparameter model systems. *Journal of Computational Physics* **26**, 1–42.

Doksum, K. and A. Samarov (1995). Nonparametric estimation of global functionals and a measure of the explanatory power of covariates in regression. *The Annals of Statistics* **23**(5), 1443–1473.

EPA (2001). Draft guidance on the development, evaluation, and application of regulatory environmental models. Environmental Protection Agency (EPA), Council for Regulatory Environmental Modeling (CREM). http://cfpub.epa.gov/crem/cremlib.cfm whitepapers.

Esty, D., M. Levy, T. Srebotnjak, A. de Sherbinin, C. Kim and B. Anderson (2006). Pilot 2006 environmental performance index. New Haven: Yale Center for Environmental Law and Policy.

Farrell, K. N. (2007). Steering into the skid: the challenge of operationalising type II error avoidance in science for policy. Paper in progress.

Fixx, J. F. (1972). *Games for the Super Intelligent*. Garden City, NY: Doubleday & Co. Inc.

Funtowicz, S. (2004). Models of science and policy: from expert demonstration to post normal science. In *ViewInternational Symposium: Uncertainty and Precaution in Environmental Management*, Copenhagen. http://upem.er.dtu.dk/programme.htm.

Funtowicz, S., M. O'Connor, S. Faucheux, G. Froger and G. Munda (1996). Emergent complexity and procedural rationality: post-normal science for sustainability. In R. Costanza, J. Martinez-Alier and O. Segura (eds), *Getting Down to Earth: Practical Applications of Ecological Economics*. Washington, DC: Island Press/ISEE, pp. 223–248.

Funtowicz, S. O. and J. R. Ravetz (1990). *Uncertainty and Quality in Science for Policy*. Dordrecht: Kluwer Academic Publishers.

Funtowicz, S. O. and J. R. Ravetz (1993). Science for the post-normal age. *Futures* **25**, 735–755.

Giona, M. and O. Paladino (1994). Bifurcation analysis and stability of controlled CSTR. *Computers and Chemical Engineering* **18**, 877–887.

Grieb, T., N. Shang, R. C. Spear, S. A. Gherini and R. A. Goldstein (1999). Examination of model uncertainty and parameter interaction in the global carbon cycling model. *Environment International* **25**, 787–803.

Grievank, A. (2000). *Evaluating Derivatives, Principles and Techniques of Algorithmic Differentiation*. SIAM.

Hankinson, S. E. Sc.D., J. E. Manson, F. E. Speizer and G. A. Colditz (eds) (2001). *Healthy Women, Healthy Lives: A Guide to Preventing Disease, from the Landmark Nurses' Health Study.* Free Press.

Hastie, T. and R. Tibshirani (1990). *Generalized Additive Models.* Chapman and Hall.

Helton, J., J. Johnson, C. Sallaberry and C. Storlie (2006). Survey of sampling-based methods for uncertainty and sensitivity analysis. *Reliability Engineering and System Safety* **91** (10–11), 1175–1209.

Heston, S. (1993). A closed-form solution for options with stochastic volatility with applications to bond and currency options. *Review of Financial Studies* **6**, 327–343.

Hodrick, T. and E. Prescott (1980). Post-war US business cycles: an empirical investigation. Manuscript, Carnegie Mellon University.

Høeg, P. (1995). *Borderliners.* Seal Books Publisher.

Hoeting, J. A., D. Madigan, A. E. Raftery and C. T. Volinsky (1999). Bayesian model averaging: a tutorial. *Statistical Science* **14**(4), 382–417.

Homma, T. and A. Saltelli (1996). Importance measures in global sensitivity analysis of nonlinear models. *Reliability Engineering and System Safety* **52**, 1–17.

Hora, S. and R. Iman (1986). A comparison of maximum/bounding and Bayesian/Monte Carlo for fault tree uncertainty analysis. Report SAND85-2839, Sandia Laboratories.

Hornberger, G. and R. Spear (1981). An approach to the preliminary analysis of environmental systems. *Journal of Environmental Management* **7**, 7–18.

Hull, J. (1997). *Options, Futures and Other Derivatives.* Prentice-Hall Inc.

Iman, R. and S. Hora (1990). A robust measure of uncertainty importance for use in fault tree system analysis. *Risk Analysis* **10**(3), 401–403.

In 't Veld, R. J. (2000). *Willingly and Knowingly.* The Hague: **LEMMA** Publishers.

IPCC (1999). IPCC expert meetings on good practice guidance and uncertainty management in national greenhouse gas inventories. Background papers. http://www.ipcc-nggip.iges.or.jp/public/gp/gpg-bgp.htm.

IPCC (2000). Good practice guidance and uncertainty management in national greenhouse gas inventories. http://www.ipcc-nggip.iges.or.jp/public/gp/gpgaum.htm.

Ishigami, T. and T. Homma (1996, December 3–5). An importance qualification technique in uncertainty analysis for computer models. *Proceedings of the isuma '90*, First International Symposium on Uncertainty Modelling and Analysis, University of Maryland.

Jakeman, A. J. and P. C. Young (1984). Recursive filtering and the inversion of ill-posed causal problems. *Utilitas Mathematica* **35**, 351–376.

Kalman, R. (1960). A new approach to linear filtering and prediction problems. *ASME Transactions, Journal of Basic Engineering* **82D**, 35–45.

Kass, R. E. and A. E. Raftery (1995). Bayes factors. *Journal of the American Statistical Association* **90**(430), 773–795.

Kennedy, P. (2007). *A Guide to Econometrics*, 5th edn. Blackwell Publishing.

Kioutsioukis, I., D. Melas, C. Zerefos and I. Ziomas (2005). Efficient sensitivity computations in 3d air quality models. *Computer Physics Communications* **167**, 23–33.

Kleijnen, J. (2007a). *DASE: Design and Analysis of Simulation Experiments.* Springer Science + Business Media (forthcoming).

Kleijnen, J. (2007b). Kriging metamodeling in simulation: a review. CentER discussion paper.

Konikov, L. and J. Bredehoeft (1992). Groundwater models cannot be validated. *Advances in Water Resources* **15**(1), 75–83.

Krzykacz-Hausmann, B. (1990). Gesellschaft fuer Reaktor Sicherheit (GRS) MbH. Technical Report GRS-A-1700, Garching.

Leamer, E. (1978). *Specification Searches: Ad Hoc Inferences with Nonexperimental Data*. John Wiley & Sons, Ltd.

Leamer, E. (1990). Let's take the con out of econometrics, and sensitivity analysis would help. In C. Granger (ed.), *Modelling Economic Series*. Oxford: Clarendon Press.

Le Dimet, F.-X., I. Navon and D. Daescu (2002). Second order information in data assimilation. *Monthly Weather Review 130*, 629–648.

l'Ecuyer, P. and T. Andres (1997). A random number generator based on the combination of four LCGs. *Mathematics and Computers in Simulation 44*, 99–107.

Li, G., J. Hu, S.-W. Wang, P. Georgopoulos, J. Schoendorf and H. Rabitz (2006). Random Sampling-High Dimensional Model Representation (RS-HDMR) and orthogonality of its different order component functions. *Journal of Physical Chemistry A 110*, 2474–2485.

Li, G., S. -W. Wang and H. Rabitz (2002). Practical approaches to construct RS-HDMR component functions. *Journal of Physical Chemistry 106*, 8721–8733.

Little, R. J. A. and D. Rubin (2002). *Statistical Analysis with Missing Data*. Hoboken, New Jersey: J. Wiley & Sons, Ltd.

McKay, M. (1996). Variance-based methods for assessing uncertainty importance in nureg-1150 analysis. Technical Report LA-UR-96-2695, 1, Los Alamos Laboratories.

Mac Lane, S. (1988). *Science, Letters 241*, 1144. *Science, Letters 242*, 1623, 1624.

Mallet, V. and B. Sportisse (2004). 3-D chemistry-transport model Polair: numerical issues, validation and automatic-differentiation strategy. *Atmospheric Chemistry and Physics Discussions 4*, 1371–1392.

Michaels, D. (2005, June). Doubt is their product. *Scientific American 292*(6).

Morbidelli, M. and A. Varma (1988). A generalized criterion for parametric sensitivity: application to thermal explosion theory. *Chemical Engineering Science 43*, 91–102.

Morris, M. D. (1991). Factorial sampling plans for preliminary computational experiments. *Technometrics 33*, 161–174.

Nardo, M., M. Saisana, A. Saltelli, S. Tarantola, A. Hoffman and E. Giovannini (2005). Handbook on constructing composite indicators: methodology and user guide. OECD statistics working paper. www.olis.oecd.org/olis/2005doc.nsf/LinkTo/std-doc (2005)3.

NIH (2002). *Framingham Heart Study: 50 Years of Research Success*. National Institutes of Health, National Heart, Lung and Blood Institute. www.nhlbi.nih.gov/about/framingham.

Oakley, J. and A. O'Hagan (2004). Probabilistic sensitivity analysis of complex models: a Bayesian approach. *Journal of the Royal Statistical Society B 66*, 751–769.

Oreskes, N., K. Shrader-Frechette and K. Belitz (1994). Verification, validation, and confirmation of numerical models in the earth sciences. *Science 263*, 641–646.

Paladino, O. and M. Ratto (2000). Robust stability and sensitivity of real controlled CSTRs. *Chemical Engineering Science 55*, 321–330.

Paladino, O., M. Ratto and P. Costa (1995). Chaos and chemical reactor models: sensitivity of dynamics on parameters uncertainty. *Chemical Engineering Science 50*, 3829–3833.

Pandis, S. N. and J. H. Seinfeld (1989). Sensitivity analysis of a chemical mechanism for aqueous phase atmospheric chemistry. *Journal of Geophysical Research 94*, 1105–1126.

Pappenberger, F., I. Iorgulescu and K. J. Beven (2006). Sensitivity analysis based on regional splits and regression trees (SARS-RT). *Environmental Modelling and Software 21*(7), 976–990.

BIBLIOGRAPHY

Pappenberger, F. and V. Stauch (2007). Estimating Sobol' sensitivity with cubic splines. *Journal of Hydrology* (submitted).

Pellegrini, L. and G. Biardi (1990). Chaotic behaviour of a controlled **CSTR**. *Computers and Chemical Engineering* **14**, 1237–1247.

Pilkey, O. H. and L. Pilkey-Jarvis (2007). *Useless Arithmetic. Why Environmental Scientists Can't Predict the Future*. New York: Columbia University Press.

Press, W. H., S. Teukolsky, W. T. Vetterling and B. P. Flannery (1997). *Numerical Recipes in Fortran 77: The Art of Scientific Computing* (Second ed.), Volume 1. New York: Cambridge University Press.

Rabitz, H. (1989). System analysis at molecular scale. *Science* **246**, 221–226.

Ratto, M., A. Pagano and P. C. Young (2007). State dependent parameter meta-modelling and sensitivity analysis. *Computer Physics Communications* **177**, 863–876.

Ratto, M., S. Tarantola, A. Saltelli and P. C. Young (2004). Accelerated estimation of sensitivity indices using State Dependent Parameter models. In K. M. Hanson and F. M. Hemez (eds), *Sensitivity Analysis of Model Output, Proceedings of the 4th International Conference on Sensitivity Analysis of Model Output (SAMO 2004)* Santa Fe, New Mexico, 8–11 March 2004, pp. 61–70. http://library.lanl.gov/ccw/samo2004/.

Ravetz, J. (2006). Models as metaphors. In B. e. a. Kasemir (ed.), *Public Participation in Sustainability Science*. Cambridge University Press.

Rosen, R. (1991). *Life Itself – a Comprehensive Inquiry into Nature, Origin, and Fabrication of Life*. Columbia University Press.

Sacks, J., W. J. Welch, T. J. Mitchell and H. P. Wynn (1989). Design and analysis of computer experiments. *Statistical Science* **4**, 409–435.

Saisana, M., A. Saltelli and S. Tarantola (2005). Uncertainty and sensitivity analysis techniques as tools for the quality assessment of composite indicators. *Journal of the Royal Statistical Society A* **168**(2), 307–323.

Saltelli, A. (2002). Making best use of model valuations to compute sensitivity indices. *Computer Physics Communications* **145**, 280–297.

Saltelli, A. (2006). Composite indicators between analysis and advocacy. *Social Indicators Research*. DOI 10.1007/s11205-006-0024-9.

Saltelli, A., T. Andres and T. Homma (1993). Sensitivity analysis of model output: an investigation of new techniques. *Computational Statistics and Data Analysis* **15**, 211–238.

Saltelli, A., K. Chan and M. Scott (eds) (2000). *Sensitivity Analysis*. Wiley Series in Probability and Statistics. New York: John Wiley & Sons, Ltd.

Saltelli, A. and J. Hjorth (1995). Uncertainty and sensitivity analyses of **OH**-initiated dimethyl sulphide (**DMS**) oxidation kinetics. *Journal of Atmospheric Chemistry* **21**, 187–221.

Saltelli, A., M. Ratto, S. Tarantola and F. Campolongo (2005). Sensitivity analysis for chemical models. *Chemical Reviews* **105**, 2811–2828.

Saltelli, A. and S. Tarantola (2002). On the relative importance of input factors in mathematical models: safety assessment for nuclear waste disposal. *Journal of the American Statistical Association* **97**, 702–709.

Saltelli, A., S. Tarantola, F. Campolongo and M. Ratto (2004). *Sensitivity Analysis in Practice: A Guide to Assessing Scientific Models*. John Wiley & Sons, Ltd.

Saltelli, A., S. Tarantola and K. Chan (1999). Quantitative model-independent method for global sensitivity analysis of model output. *Technometrics* **41**(1), 39–56.

Schulze, N. and D. Baur (2006, May). Economic impact study on clearing and settlement – annex ii. Technical report, European Commission, DG-JRC. available at http://ec.europa.eu/internal_market/financial-markets/clearing/index_en.htm.

Schweppe, F. (1965). Evaluation of likelihood functions for Gaussian signals. *IEEE Transactions on Information Theory* 11, 61–70.

SIMLAB (2007). Ver 3.0. http://sensitivity-analysis.jrc.cec.eu.int/, European Commission, Joint Research Centre, Ispra.

Sobol', I. M. (1967). On the distribution of points in a cube and the approximate evaluation of integrals. *USSR Computational Mathematics and Mathematical Physics* 7, 86–112.

Sobol', I. M. (1976). Uniformly distributed sequences with additional uniformity properties. *USSR Computational Mathematics and Mathematical Physics* 16(5), 236–242.

Sobol', I. M. (1990). Sensitivity estimates for nonlinear mathematical models. *Matematicheskoe Modelirovanie* 2, 112–118. in Russian, translated in English in Sobol' (1993).

Sobol', I. M. (1993). Sensitivity analysis for non-linear mathematical models. *Mathematical Modelling and Computational Experiment* 1, 407–414. English translation of Russian original paper Sobol' (1990).

Sobol', I. M. (1996). On 'freezing' unessential variables. *Vestnik Moskovskogo Universiteta, Serija Matematika* 6, 92–94.

Sobol', I. M., S. Tarantola, D. Gatelli, S. Kucherenko and W. Mauntz (2007). Estimating the approximation error when fixing unessential factors in global sensitivity analysis. *Reliability Engineering and System Safety* 92, 957–960.

Spear, R. (1997). Large simulation models: calibration, uniqueness and goodness of fit. *Environmental Modelling and Software* 12, 219–228.

Spear, R., T. Grieb and N. Shang (1994). Factor uncertainty and interaction in complex environmental models. *Water Resources Research* 30, 3159–3169.

Storlie, C. and J. Helton (2008). Multiple predictor smoothing methods for sensitivity analysis: description of techniques. *Reliability Engineering and System Safety* 93, 28–54.

Taleb, N. N., (2007), The Black Swan. *Penguin*. London.

Tarantola, S., D. Gatelli and T. Mara (2006). Random balance designs for the estimation of first order global sensitivity indices. *Reliability Engineering and System Safety* 91(6), 717–727.

Tarantola, S., R. Pastorelli, M. G. Beghi and C. E. Bottani (2000). A dataless pre-calibration analysis in solid state physics. In A. Saltelli, K. Chan and M. Scott (eds), *Sensitivity Analysis*, pp. 311–327. John Wiley & Sons, Ltd.

Turanyi, T. (1990). Sensitivity analysis of complex kinetic systems. Tools and applications. *Journal of Mathematical Chemistry* 5, 203–248.

Van der Sluijs, J. P. (2002). A way out of the credibility crisis of models used in integrated environmental assessment. *Futures* 34, 133–146.

Van der Sluijs, J., Craye, M., Funtowicz, S., Kloprogge, P., Ravetz, J. and Risbey J. (2005). Experiences with the NUSAP system for multidimensional uncertainty assessment in model based foresight studies. *Water Science and Technology* 52(6), 133–144.

Varma, A., M. Morbidelli and H. Wu (1999). *Parametric Sensitivity in Chemical Systems*. Cambridge University Press.

Vuilleumier, L., R. Harley and N. Brown (1997). First- and second-order sensitivity analysis of a photochemically reactive system (a Green's function approach). *Environmental Science and Technology* 31, 1206–1217.

BIBLIOGRAPHY

Wichman, B. and I. Hill (1982). Algorithm **AS** 183: an efficient and portable pseudo-random number generator. *Applied Statistics* **31**, 188–190.

Young, P. C. (1999a). Data-based mechanistic modelling, generalised sensitivity and dominant mode analysis. *Computer Physics Communications* **117**, 113–129.

Young, P. C. (1999b). Nonstationary time series analysis and forecasting. *Progress in Environmental Science* **1**, 3–48.

Young, P. C. (2000). Stochastic, dynamic modelling and signal processing: time variable and state dependent parameter estimation. In W. J. Fitzgerald. *et al.* (eds), *Nonlinear and Nonstationary Signal Processing*, pp. 74–114. Cambridge: Cambridge University Press.

Young, P. C. (2001). The identification and estimation of nonlinear stochastic systems. In A. I. Mees (ed.), *Nonlinear Dynamics and Statistics*. Boston: Birkhäuser.

Young, P. C. (2002). Advances in real-time flood forecasting. *Philosophical Trans. Royal Society, Physical and Engineering Sciences* **360**, 1433–1450.

Young, P. C., P. McKenna, and J. Bruun (2001). Identification of nonlinear stochastic systems by state dependent parameter estimation. *International Journal of Control* **74**, 1837–1857.

Young, P. C. and C. N. Ng (1989). Variance intervention. *Journal of Forecasting* **8**, 399–416.

Young, P. C., S. Parkinson and M. Lees (1996). Simplicity out of complexity: Occam's razor revisited. *Journal of Applied Statistics* **23**, 165–210.

Young, P. C. and D. J. Pedregal (1996). Recursive fixed interval smoothing and the evaluation of Lidar measurements. *Environmetrics* **7**, 417–427.

Young, P. C. and D. J. Pedregal (1999). Recursive and en-bloc approaches to signal extraction. *Journal of Applied Statistics* **26**, 103–128.

Young, P. C., R. C. Spear and G. M. Hornberger (1978). Modeling badly defined systems: some further thoughts. In *Proceedings SIMSIG Conference*, Canberra, pp. 24–32.

Index

Analytical g-function 123–7
ANOVA-HDMR decomposition 162, 213, 221, 274
Approximating functions 212–13
Asymptotic curves 57

Batch reactor 262–70
    see also Thermal runaway analysis
Bayesian model averaging 8–9
Bootstrapp 7–8

Chemical reactor 262–70
    see also Thermal runaway analysis
Clusters 58, 83
Composite index, see Composite indicator
Composite indicator 240–53
    aggregation methods 239, 241, 250–2
    arithmetic v. geometric approach 239–40
    baseline ranking 245–6
    composite scores 244–5
    and country rankings 242–3
    cumulative shift 245, 246–7, 248
    and elementary effects (EE) 239, 246–7, 252
    input factors
        impact 243–5
        number reduction 246
    and interactions 247
    mapping 247–50
    Monte Carlo Filtering (MCF) 248–50, 252

behavioural/nonbehavioural subsets 248, 249–50
    robustness 241
    Smirnov statistics 250–1
    uncertainty analysis 242–3, 244
    validity 241
    variance-based methods 239, 246–7, 252
    weights 247
Conditional expectation 160–1
Conditional variances 20–2
Continuous stirred tank reactor (CSTR)
    as dynamical system 202
    heat balance 201
    mass balance 200, 201
    stability conditions analysed 202–6
        Hopf bifurcation locus 203–5
        robustness check 204–6
        Smirnov analysis 204, 205–6
        uncertainties 206
Correlation ratio 213
Cost of analysis 17
'Counterfeit Coin Puzzle' 90–1
Cubic polynomial spline 218

Data mining 54
Decomposition 160, 161–2
    ANOVA-HDMR 162, 213, 221, 274
    and risk 157
    variance-based methods 19–20, 160, 161–2
Derivatives
    advantages/disadvantages 11–12
    as basis of sensitivity analysis 11

Global Sensitivity Analysis. The Primer A. Saltelli, M. Ratto, T. Andres, F. Campolongo, J. Cariboni, D. Gatelli, M. Saisana and S. Tarantola © 2008 John Wiley & Sons, Ltd

Derivatives (*Continued*)
  compared with scatterplots 14–15
  normalization 15–16
Deterministic models 157
Deterministic regularization (DR) 218
Discontinuous functional forms 57–8
Discrepancy 83, 84
Distribution of points 59–60

Elementary effects (EE) 109–54
  advantages 127–8, 274, 275
  analytical *g*-function 123–7
  composite indicator application
    239, 246–7
  defined 110–11, 121
  and groups 121–2, 128
  role of delta (Δ) 120–1
  sampling strategy 112–16
    optimization 115–16
  sensitivity measure computation
    110–11, 116–22
    factor fixing 125
    practical example 123–7
    standard deviation 110, 111, 117
  test defined 38–9
Endpoints 60, 62, 63–4
Errors 15, 166
  standard error and uncertainty 59
  see also type I errors, type
    II errors, type III errors
Experimental design 35, 53–107
  group sampling 89–96
  and multiple parameters 64–89
  and single parameter 55–64

Factor fixing (FF) 33–4, 125, 156
Factor mapping (FM) 39, 40, 156–7,
  183–236
Factor Prioritization (FP) 24–5, 156
Factorial design, *see* Fractional factorial
  (FF) sampling
Factors 5–6, 7
  distribution 10, 25
  in experimental design 54
  groups or sets 36–7
  identification 35
  independence 17
  influence 21, 24, 26, 27, 258–60
    and choice of technique 272,
      274–5
    and jumps in pricing options 257,
      258–60, 261, 262

and Monte Carlo filtering 209–10
  nonindependent 41
  selection 9–10
First-order effect 21
First-order sensitivity index, *see*
  Sensitivity index, first-order
'Fitness for purpose' 4–5, 10, 43
Fourier Amplitude Sensitivity Test
  (FAST) 159, 167
Fractional factorial (FF) sampling
  71–6, 89, 274–5
  Hadamard matrix 73–4
  and LH sampling combined
    82, 106
  main effect (ME) of parameters 75
  and simulations 72
Framingham Heart Study 53
Fussell–Vesely measure 157

g-function 123–7
Gaps 58, 83
Gaussian emulators 214–15
Generalized Random Walk (GRW)
  222
Group sampling 89–96
  number 92
  parameters
    allocation 92, 93
    influential 93, 94–6
    noninfluential 93–4
  sign variables 95, 96
  and simulations required 89
  stepwise analysis 95–6
  supersaturated designs 89–90
Groups 36–7, 89–96
  and elementary effects method
    109–10
  and scatterplots 15
  see also Group sampling

Haar wavelet 216–18
Hadamard matrix 73–4
Halton sequence 84–6
  radical inverse transform 86
Health studies 53–4
High-dimensional model representation
  (HDMR) 160, 227, 228, 236
  estimating 214–24
  smoothing techniques (Haar wavelet)
    216–18
  spline smoothing 218–21

INDEX

state-dependent regressions 221–4, 227, 228
see also ANOVA-HDMR decomposition
Hodrick–Prescott (HP) filter 218–21, 223, 226
Hopf bifurcation locus 203–5

Index/indices, *see* Sensitivity index
Infection dynamics
  model 169–74
    input factors 209–10
    and Monte Carlo filtering 209–11
    and Random Balance Design (RBD) 174
    and sensitivity index 170–1
    and uncertainty analysis 171
    and variance-based methods 169–74
Input factors, *see* Factors; Parameters
Integrated Random Walk (IRW) 222, 223
Interactions 30, 31, 268, 269, 272
  definition 161
  and metamodelling 274
  in variance-based method 161–2

Kennedy, Peter 42
Kernel regression methods 213–14
Kriging metamodels 214

Latin hypercube (LH) sampling 76–80, 89, 103–5
Leamer, Edward E. 9–10
Least-square computation 17–18, 66
Linear models 22–3
  and experimental design 65–6
    least-squares solution 17–18, 66
    random samples 66
    regression analysis 66
    one-at-a-time (OAT) sampling 69
Linear polynomials 57
Linear regression 17–19
Log-transformation 234
Low-discrepancy sequence 83–9
  defined 83
  Halton sequence 84–6
  see also Quasi-random sampling

Macroeconomic model 206–9
  backward-looking/forward-looking components 207

Phillips curve 206–7
stability conditions 207, 208, 209
stable/unstable behaviour 207–9
Main effect 75
Mapping
  Environmental Performance Index 247–50
  factor mapping (FM) 39, 40, 156–7, 183–236
  log-transformed functions 234
Mean, as model output 157–8
Mean estimates, stratified sampling 61–4
Metamodelling 43, 183–236, 274, 275
  approximating functions 212–13
  interpolating 214–15
    Gaussian emulators 214–15
    kriging metamodels 214
  kernel regression methods 213–14
  methods summarized 212
  and Monte Carlo Filtering (MCF) 184–211, 235
  purposes 215
  smoothing techniques 214–20
Mirror points 69
Model approximation 212–35
Model coefficient of determination 19
'Model-free' approach 20
Models 1–10
  additive 23, 25
  characteristics 277–8
  deterministic 157
  functions 4
  inputs, *see* Factors; Parameters
  linear, *see* Linear models
  nonadditive 23, 25–9
  nonlinear 19, 23
  parameter estimation 6–10
  parsimonious 43
  relevance 34
  Rosen's 2
  simplification 33–4, 35
  and simulation requirements 89
  types 5
  unstable 128
Modulus incremental ratios 45

Monte Carlo Filtering (MCF) 39–40, 41, 184–211, 275
  behavioural/nonbehavioural subsets 39, 40, 184–6, 248, 249–50
  bidimensional projections 186–7
  and composite indicator 239, 248–50, 252
  continuous stirred tank reactor (CSTR) 200–6
  definition 184, 248
  implementation 185–7
  infection dynamics model 209–11
  macroeconomic model 206–9
  and metamodelling 184–211, 235
  parameter importance 185–6
  Regionalized Sensitivity Analysis (RSA) 184–5, 187–8
  Smirnov test/analysis 185–6, 187–8, 204, 205–6
  stability analysis 200–11
  Tree-Structured Density Estimation (TSDE) technique 188
Monte Carlo method 6–7, 13, 16–20
  and first-order sensitivity measures 25–6
  and sensitivity index computation 164–7
    error estimates 166
Multiple parameters 64–89
Multivariate stratified sampling 80–2
  fractional factorial (FF) approach 81
  LH and FF approaches combined 82
  sample point generation 80–1

Noise Variance Ratio (NVR) hyperparameter 222, 223
Nonparametric R-squared 213
Normalization 15–16, 56
Null hypothesis 90

Oakley–O'Hagan function 129, 130, 145–6
One-at-a-time (OAT) sampling 66–9, 89, 109
  balancing 67–9
  parameter changes 69, 75
Orthogonal arrays 79–80, 106

Piecewise linear fit 62, 63
Post-Normal Science (PNS) 4, 277

Pricing options 253–62
  arbitrage-free prices 254
  Carr and Madan pricing method 256–7
  controllable/uncontrollable factors 257, 258–60, 261
  Cox–Ingersoll–Ross process 256
  Heston model 255–7
  input factors 257, 258–60
  jump parameters 258–60, 261, 262
  method selection 258
  risk 254–5
  strike price 254, 258, 260, 261
  uncertainty 255, 261
  volatility 262
Pseudo-random generator 83, 100–1

Quadratic polynomials 57
Quantiles 119, 140
Quasi-random numbers 274
Quasi-random sampling 83–9
  and sample size 89
  Sobol' LP_τ sequence 87
  testing 86–7
  uncertainty estimates 89
  see also Low-discrepancy sequence

Radical inverse transform 86
Random Balance Design (RBD) 167–9, 274
  advantages/disadvantages 168–9, 236
  and infection dynamics 169–74
  procedure 167–8
Random samples 58–9, 66
  pseudo-random generator 83, 100–1
Regionalized Sensitivity Analysis (RSA) 184–5, 187–8
  limitations 188
Regression coefficients 18
Regression methods 17–18, 37, 66, 213–14, 274
Regularization 218
Residuals 66
Resolution III 74
Resolution IV 74, 103, 107
Risk decomposition 157
Risk reduction worth 157
Rosen, R. 2

INDEX

Saltelli's method 164–7, 236, 272, 274
Sample matrices 274
Sampling strategy
  and elementary effects 112–16
Scatterplots 13–14, 15
  and derivatives 14–15
  point interpolation 37
  shape or pattern 21
  slicing 21–2, 23
  smoothing 216–18
Sensitivity analysis
  cost of 17
  definition 1
  global v. local 11–12, 35–6
  graphical presentation 271–2
  methods 10–40
  practical applications 237–75
  problems 41–2
  purposes 11, 34–6
  set-up considerations 237
Sensitivity measure 21
  applied to linear model 22–3
  computation 164–9
    acceleration 38
    FAST method 167
    from smoothed estimates 224–9
    Haar wavelet smoothing 225–6
    method choice 235–6
    Monte Carlo procedure 164–7
    RBD 167–9
    Saltelli's method 164–7, 236
    spline smoothing (HP filter) 226
  computational cost (CPU time) 272
  defined 21
  for the elementary effects method 110–11, 116–27
  first-order 21, 24, 25–6, 28, 30, 37
    methods compared 173
    Monte Carlo computation 164, 165
    suitability 275
    and variance 161
  higher order 29–31
  and infection dynamics 170–1
  and Monte Carlo method 25–6
  properties 166–7
  second order 30
  variance-based 258, 261
Sensitivity measures
Sensitivity pattern 33

Sensitivity tests
  settings 155–7
    Factor Fixing (FF) 156
    Factor Mapping (FM) 156–7
    Factor Prioritization (FP) 24, 156
    Variance Cutting (VC) 156
Sets, see Groups
Settings 10–40
  definition 24
Simulations 89
  and fractional factorial (FF) sampling 72
  group sampling 89
  and models 89
  number determined by parameters 92, 102–3
Slicing 21–2, 23
Smirnov test/analysis 185–6, 187–8, 204, 205–6, 250–1
Smoothing techniques 214–20, 235
  examples 224–9
  SDR techniques 221–4, 226–7
  spline smoothing 218–21, 223
  using Haar wavelet 216–18, 225–6
Sobol', I. M. 160
Sobol' procedure 87, 266–7, 268, 272
Spline smoothing 218–21, 223
  Hodrick–Prescott (HP) filter 218–21, 226
  'trend' 220
Stability analysis
  continuous stirred tank reactor (CSTR) 200–6
  infection dynamics model 209–11
  macroeconomic model 206–9
Standard error 59
Standardized regression coefficients (SRCs) 18, 26, 274, 275
State-dependent parameter (SDP)
  and HDMR 221–2
State-dependent regression (SDR)
  approach 214, 221–4, 226–7, 235, 236
  advantages/disadvantages 223, 236
  and HDMR 227, 228
Stratified sampling 59–61
  mean estimates 61–4
  multivariate 80–2
  point distribution 59–60
  variance estimates 61–4
Supersaturated designs 89–90

Taylor rules 201–2
Thermal runaway analysis 263–70
  ANOVA-HDMR terms 268, 269, 270
  interactions 268, 269
  metamodelling 266
  method choice 266
  procedure 266–9
  runaway/nonrunaway conditions 264, 265
  Semenov number 265
  Sobol' procedure 266–7, 268
  and state-dependent parameter (SDP) 266, 268
  temperature behaviour 266–9
  uncertainty distributions 264–6
Total effects 112, 162–3, 173, 275
  estimation 164, 165, 236
  and sensitivity pattern 33
  terms 31–3
Total indices, see Total effects
Total sensitivity index
  definition 112
  see also Total effects
Total System Performance Assessment (TSPA) 278
Tree-Structured Density Estimation (TSDE) technique 188
Type I errors 15, 177
Type II errors 42, 177, 277
  definition 15
  protection against 35, 36
Type III errors 15, 42, 278

Uncertainty 1, 3–7, 35, 157–8
  and chemical reactor 262
  and composite indicator scores 242–3, 244
  and groups 36–7
  graphical presentation 271
  and infection dynamics 171
  input factors 8
  and jumps in pricing options 255
  quantification 158
  and standard error 59
  'Uncertainty importance' 159
Uncertainty–sensitivity plot 270–2
  purposes 271–2

Variables, see Factors; Parameters
Variance-based methods 37–8, 155–82
  advantages 157–8
  and composite indicator 239, 246–7, 252
  decomposition 19–20, 160, 161
  disadvantages 158, 174
  first-order variance term 159
  Fourier Amplitude Sensitivity Test (FAST) 167
  historical aspects 159–61
  infection dynamics model 169–74
  and interaction effects 161–2
  Random Balance Designs (RBD) 167–9
  sensitivity index computation 164–9
  settings 155–7
  total effects 162–3
  uncertainty measurement 158
Variance cutting (VC) setting 45, 156
Variance estimates 61–4

Yucca Mountain repository for radioactive waste disposal 278